Biodeconstruction

SUNY series in Contemporary Continental Philosophy

Dennis J. Schmidt, editor

Biodeconstruction
Jacques Derrida and the Life Sciences

By
Francesco Vitale

Translated by
Mauro Senatore

Published by State University of New York Press, Albany

© 2018 State University of New York

All rights reserved

Printed in the United States of America

No part of this book may be used or reproduced in any manner whatsoever without written permission. No part of this book may be stored in a retrieval system or transmitted in any form or by any means including electronic, electrostatic, magnetic tape, mechanical, photocopying, recording, or otherwise without the prior permission in writing of the publisher.

For information, contact State University of New York Press, Albany, NY
www.sunypress.edu

Library of Congress Cataloging-in-Publication Data

Names: Vitale, Francesco, author | Senatore, Mauro, translator
Title: Biodeconstruction / Francesco Vitale, author.
Description: Albany : State University of New York Press, [2018] | Series: SUNY series in Contemporary Continental Philosophy | Includes bibliographical references and index.
Identifiers: ISBN 9781438468853 (hardcover : alk. paper) | ISBN 9781438468860 (ebook) | ISBN 9781438468846 (paperback : alk. paper)
Further information is available at the Library of Congress.

10 9 8 7 6 5 4 3 2 1

Contents

Acknowledgments		vii
Introduction		1
I.	Toward Biodeconstruction	7
II.	Between Life and Death: Différance	29
III.	The Absolute Programme	53
IV.	The Text and the Living	103
V.	Between Life and Death: The Bond	127
VI.	Beyond Life and Death: Autoimmunity	167
VII.	Living On: The Arche-performative	185
Notes		201
Bibliography		237
Index		243

Acknowledgments

Over the years I have contracted debts of gratitude that I cannot honor with a mere "thanks" at the opening of a book. However, I must thank Geoffrey Bennington and Rodolphe Gasché for their support and their invaluable comments on the draft of this book. But, above all, for their scholarship and friendship, which have enriched my life in an inestimable way.

I would also like to thank everyone who gave me the opportunity to present and discuss my work as it developed. In particular, Bernard Stiegler, for inviting me to give a talk at his Summer School at Epinéuil in 2013 and at the conference *General Organology* at the University of Kent in November 2014, and Michael Naas, for asking me to present my work at the Collegium Phenomenologicum in July 2015. I thank everyone who participated in the formal and informal discussions that followed these presentations and gave me the opportunity to test my readings. In particular, Giovanna Borradori, Danielle Cohen-Levinas, Silvano Facioni, Raoul Frauenfelder, Juan Manuel Garrido, Marc Goldschmit, Martin Hägglund, David E. Johnson, Peggy Kamuf, Ronald Mendoza de Jesus, Thomas Clément Mercier, Gerald Moore, and Paolo Vignola.

Finally, I thank Mme. Marguerite Derrida for allowing me to quote Jacques Derrida's unedited seminar *La vie la mort* and Geoffrey Bennington for revising the English translation of the passages quoted from this seminar.

Thanks to everyone who will allow this book to survive the time of a reading.

Introduction

Learning to Live Finally, Jacques Derrida's last interview, released when death was imminent, suggests that we take a step back and reread his oeuvre on the track of life:

> As I recalled earlier, already from the beginning, and well before the experiences of surviving [*survivance*] that are at the moment mine, I maintained that survival is an originary concept that constitutes the very structure of what we call existence, *Dasein*, if you will. We are structurally survivors, marked by this structure of the trace and of the testament. But, having said that, I would not want to encourage an interpretation that situates surviving on the side of death and the past rather than life and the future. No, deconstruction is always on the side of the *yes*, on the side of the affirmation of life. Everything I say—at least from *Pas* (in *Parages*) on—about survival as a complication of the opposition life/death proceeds in me from an unconditional affirmation of life.[1]

Along this reverse path, we encounter autoimmunity and the religious, the community and the political; the animal and the bestial associated with sovereignty; survival and testimony, Blanchot and literature. However, to grasp the sense of these apparently recent traces, we shall go further back and shed light on a more or less explicit engagement with life sciences (paleontology, ethology, and, above all, biology and the theory of evolution) since the very first steps of deconstruction. We shall consider the investigation of *life* not only an issue of deconstruction but the latter's very matrix; we shall think *différance* as the irreducible and structural condition of the life of the living, and thus *trace* and *text* as the structures of the organization of life (from the most elementary forms to the organization of the psychic system of the human being, to the formation of the ideal objectivities that structure life and institutions in our cultural habitat). I put this hypothesis to the test through the groundbreaking reading of the unedited seminar *La vie la mort*, which Derrida taught in 1975. The first part of the seminar is dedicated to biology and, in particular, to the biologist François Jacob, who was awarded the Nobel Prize for medicine in 1965, alongside Jacques Monod and André Lwoff, and who is the author of *The Logic of Life* (1970). In this part of the seminar, according to a hypothesis already advanced in *Of Grammatology* (1967), Derrida takes into examination the heuristic bearing of the scriptural model imported from cybernetics and adopted in biology to account for the genesis and structure of the living. In particular, he aims to verify the possible congruency of this model with the notion of "general textuality" formalized in *Of Grammatology*, in view of a deconstructive elaboration of the relationship life/death, traditionally understood as a reciprocally exclusive opposition. It is worth remarking that Derrida never abandons this hypothesis as attested in *For What Tomorrow* (2001), where it is recalled as the framework for the question of animality:

> Beginning with *Of Grammatology*, the elaboration of a new concept of the *trace* had to be extended to the entire field of the living, or rather to the life/death relation, beyond the anthropological limits of "spoken" language (or "written" language, in the ordinary sense), beyond the phonocentrism or the logocentrism that always trusts in a simple and oppositional limit between Man and the Animal. At the time I stressed that the "concepts of writing, trace, gramma or grapheme" exceeded the opposition "human/nonhuman."[2]

In the seminar, this project goes alongside a deconstruction of the philosophies of life elaborated on the basis of the metaphysics of presence and exemplarily represented by Hegel's philosophy. From this perspective, the seminar allows us to go back to *Glas* (1974) as a moment of this deconstruction of the traditional philosophy of life, in view of a differential conception of life (death). The seminar also allows us to go further back to *Husserl's Origin of Geometry: An Introduction* (1962), in the wake of this deconstructive elaboration of the question of life and of the necessary engagement with the life sciences.

The first chapter of this book is thus dedicated to the reconstruction of the path that Derrida followed at the beginning of the deconstructive adventure in view of what I call *biodeconstruction*, especially through "Freud and the Scene of Writing" and *Of Grammatology*. It is precisely through the biological and neurophysiological hypotheses formulated by Freud in the *Project for a Scientific Psychology* (1895) that Derrida breaches the way that leads him to engage with the life sciences and, above all, with the evolutionistic perspective that Leroi-Gourhan adopted in *Gesture and Speech* (1965) to go back to the prehistorical origins of the invention of technics and, in particular, of the technical devices of conservation and exteriorization of memory.

The central part of the book, the most conspicuous one, is dedicated to the analysis of the seminar *La vie la mort*, to the examination of Jacob's biology (chapters II–IV) and to Derrida's interpretation of Freud's *Beyond the Pleasure Principle*, specifically the biological speculation through which Freud aims to justify the hypothesis of the death drive as the originary tendency of the living (chapter V). This section of the seminar is further developed in "To Speculate—On 'Freud,'" published in *The Post Card* (1980). In the central part of the book I highlight the irreducible differential conditions that structure the life (death) of the living as a text, as a weave of traces and their implications for the constitution of the psychical individuality understood as the emergence of the living. Furthermore, I show how and why the effects of these structural biological conditions are propagated beyond the "natural" living, through psychic individuality, and also how they affect the "cultural" products of the living. This allows us to suspend the validity of the nature/culture opposition and thus also the opposition between the life sciences and the humanities, in view of a different (differential) articulation of these terms that the tradition imposed on us to think as opposed.

In the sixth chapter I aim to demonstrate that the introduction of the autoimmunitarian lexicon in Derrida's oeuvre from the 1990s consists in a further development of the findings of the seminar *La vie la mort*. My argument is that, to understand the bearing of this lexicon and its extension to the religious, the political, democracy and sovereignty, we should go back to the biological sources of autoimmunity, on which Derrida draws more or less explicitly and, in particular, to the theory of cellular suicide, namely, apoptosis. The irreducible co-implication of life and death structures the living in such a way that the living must relate to the other in order to be itself, but, in so doing, it must destroy its own immunitarian defenses, that is, it must suppress the immunitarian defenses of the organs that preside over the relation to alterity in view of the survival in the environment and of reproduction (brain, eyes, and female uterus).

In the last chapter I suspend the focus on biology and verify the outcome of *Biodeconstruction*: the différance life/death makes us think life as survival and survival as the condition of the testamentary and thus testimonial structure that affects the production of traces, the writing of the living, beyond the supposed biological and natural limits of the living, up to the spectral conditions of the constitution of the ideality. This part, titled "Living On: The Arche-performative," has already been published in the collected volume *Performative after Deconstruction* (2013), edited by Mauro Senatore. Here I refer to the seminar *La vie la mort* for the first time. I had to write it again in the light of the deeper analyses I have developed over the last few years, which led me to the completion of *Biodeconstruction*. But this is somehow what I have done, so long as *Biodeconstruction* may be read as the rewriting of the essay "Living On: The Arche-performative," a rewriting necessary to justify the latter's theses, as much as "Living On: The Arche-performative" may be read as the matrix or the seminal trace from which *Biodeconstruction* has been developing, by differentiation and through successive stratifications, as the elaboration of the survival I attest to be through the traces that constitute this book.

A few words about the method. In a note in "From Restricted to General Economy," Derrida describes precisely the deconstructive method adopted in his reading of Bataille, but which is valid for every text, as follows:

> Like every discourse, like Hegel's, Bataille's discourse has the form of a structure of interpretations. Each proposition, which is already interpretive in nature, can be interpreted by another proposition. Therefore, if we proceed prudently and all the while remain in Bataille's text, we can detach an interpretation from its reinterpretation and submit it to another interpretation bound to other propositions of the system. Which, without interrupting general systematicity, amounts to recognizing the strong and weak moments in the interpretation of a body of thought by itself, these differences of force keeping to the strategic necessity of finite discourse. Naturally our own interpretive reading has attempted to pass through what we have interpreted as the major moments, and has done so in order to bind them together. This "method"—which we name thus within the closure of knowl-

edge—is justified by what we are writing here, in Bataille's wake, about the suspension of the epoch of meaning and truth. Which neither frees nor prohibits us from determining the rules of force and of weakness: which are always a function of: (1) the distance from the moment of sovereignty; (2) the misconstruing of the rigorous norms of knowledge. The greatest force is the force of a writing which, in the most audacious transgression, continues to maintain and to acknowledge the necessity of the system of prohibitions (knowledge, science, philosophy, work, history, etc.). Writing is always traced between these two sides of the limit.[3]

Undoing the textual knots that bind a given system of interpretation to a certain order of knowledge and tying together again, in a different way, the propositions that constitute that system: this is the work of deconstruction. A minute and patient work that can be imperceptible but can also produce irreducible effects of destructuration in the field in which it intervenes, a field that will never be the same. Hence, to account for the work of deconstruction, we should follow with the same patience the process of destructuration of the systems of interpretation in which it intervenes in order to isolate the moments in which the graft of the deconstructive interpretation displaces the sense of the texts interpreted while remaining intimately adherent to them. For this reason, to let the sense of the deconstructive reading/writing emerge and to follow the latter's elaboration, I thought it is necessary to recur to long quotations from Derrida's texts and from those interpreted by Derrida, in particular, in order to reconstruct step by step the close engagement with Jacob and Freud, because the step of deconstruction does not come to a halt in the presence of a sentence or thesis but survives in the network of references it interweaves.

I

Toward Biodeconstruction

Let me take a little step back. I begin with the genesis of the notion of arche-writing and thus with Edmund Husserl's *"Origin of Geometry": An Introduction*.[1] It is well known that Derrida introduced the notion of arche-writing in *Speech and Phenomena* to account for the effects of the deconstruction of Husserl's "living present" and, thus, to reformulate the dynamic of retention (and of the constitution of memory) as untied from what Husserl calls the "principle of principles" of phenomenology. However, in *Edmund Husserl's "Origin of Geometry,"* Derrida had already prepared the ground for this work of deconstruction by drawing the latter's critical threshold: the necessity—from the genetic perspective of *Edmund Husserl's Origin of Geometry*—of recurring to writing in order to describe the conditions of the constitution of ideal objects. For Husserl, only the possibility of writing grants the conservation and transmission of ideal objects as written signs that are not immediately related to the living present they refer to and, thus, do not depend on the actual present of their alleged, originary production. For Derrida, the structural dynamic of writing does not concern merely the conservation and transmission of ideal objects but also, more radically, the possibility of sense in general for a consciousness in general. The retention of the trace of experience in the individual consciousness is already affected by an irreducible detachment from the immediate and living present of intuition, because only this detachment allows consciousness to recognize the "same" in a reference to come, which is absolutely other than the "living present" of its alleged, originary production:

> Before the "same" is recognized and communicated among several individuals, it is recognized and communicated within the individual consciousness: after quick and transitory evidence, after a finite and passive retention vanishes, its sense can be re-produced as the "same" in the act of recollection; its sense has not returned to nothingness. In this *coincidence of identity, ideality* is announced as such and in general in an egological subject. . . . Thus, before being the ideality of an identical object for other subjects, sense is this ideality for *other* moments of the same subject in a certain way, therefore, intersubjectivity is first the non-empirical relation of Ego to Ego, of my present to other presents as such; i.e., as others and as presents (as past presents). Intersubjectivity is the relation of an absolute origin to other absolute origins, which are always my own, despite their radical alterity. Thanks to this circulation of primordial absolutes, the *same* thing can be thought through absolutely other moments and acts.[2]

Therefore, in Edmund Husserl's "Origin of Geometry," Derrida discovers and formalizes the law of iterability as the irreducible condition of the retentional trace and, thus, of sense in general. There is no "living present" for consciousness, which is constituted by means of a certain writing *avant la lettre* that, precisely for this reason, Derrida shall call "arche-writing" later in *Speech and Phenomena*:

> The living present springs forth out of its nonidentity with itself and from the possibility of the retentional trace. It is always already a trace. This trace cannot be thought out on the basis of a simple present whose life would be within itself; the self of the living present is primordially [*originairement*] a trace. The trace is not an attribute; we cannot say that the self of the living present "primordially is" it ["*l'est originairement*"]. Being-primordial [*l'être-originaire*] must be thought on the basis of the trace, and not the reverse. This arche-writing is at work at the origin of the sense.[3]

This should be well known. But there is a passage in Edmund Husserl's "Origin of Geometry" which is perhaps less known and yet still decisive from our perspective. It is a footnote in which Derrida affirms that Husserl cannot take the step toward retention as a writing *avant la lettre* because it would require that the very limits of phenomenology, which establishes itself for an already constituted consciousness, be transgressed:

> The passage from passive retention to memory or to activity of recollection, a passage which "produces" ideality and pure objectivity as such and makes other absolute origins appear as such, is always described by Husserl as an already given essential possibility, as a structural ability whose source is not made a problem. Perhaps this source is not questioned by phenomenology because it is confused with the possibility of phenomenology itself. In "its factuality," this passage is also the one from the lower forms of Nature and life to consciousness [*Dans "sa facticité" ce passage est aussi celui des formes inférieures de la nature et de la vie à la conscience*]. It can be also the thematic site of what today is called "overcoming." Here phenomenology would be "overcome" or completed in an interpretative philosophy.⁴

In order to conceive of the genesis and structure of retention as arche-writing, it is necessary to overcome the limits of phenomenology; it is necessary to carry on the phenomenological investigation toward the genesis of consciousness, to the point of focusing on the natural and, thus, biological condition of the emergence of consciousness itself. As Derrida points out, "this passage is also the one from the lower forms of Nature and life to consciousness." It is worth remarking that this is not an occasional reference that is confined to a footnote. Derrida recalls the necessity of the overcoming of phenomenology in the passage from *Speech and Phenomena* in which he goes back to the problem addressed in *Edmund Husserl's "Origin of Geometry"*: iterability as the irreducible condition of the retentional trace. Furthermore, the passage shows that, for Derrida, the emergence of consciousness must be traced throughout the evolution of the living—here designated as "history of life"—that dictates the conditions of possibility of that emergence:

> Without reducing the abyss which may indeed separate retention from re-presentation, without hiding the fact that the problem of their relationship is none other than that of the history of "life" and of life's becoming conscious, we should be able to say

a priori that their common root—the possibility of re-petition in its most general form, that is, the constitution of a trace in the most universal sense—is a possibility which not only must inhabit the pure actuality of the now but must constitute it through the very movement of difference it introduces. Such a trace is—if we can employ this language without immediately contradicting it or crossing it out as we proceed—more "primordial" than what is phenomenologically primordial.⁵

The traces of this movement across phenomenology and, therefore, this investigation of the natural genesis of consciousness as a specific emergence within the history of life, that is, of evolution, are already evident in "Freud and the Scene of Writing" (1966) and in *Of Grammatology* (1967).⁶

The Biological Genesis of Arche-writing

In "Freud and the Scene of Writing," Derrida focuses on the necessity that compels Freud to recur to the metaphor of writing and to scriptural devices that are more and more complex, up to "Note on the Mystic Writing-Pad" (1925), in view of describing and explaining the genesis and structure of the psychic system. However, Derrida's concern is analyzing the effects that this metaphorical escalation has on Freud's conception of the psychical system—taking for granted what writing and text properly are—as well as the effects that the recourse to the metaphor has on the traditional notions of "writing" and "text":

> We shall let our reading be guided by this metaphoric investment. It will eventually invade the entirety of the psyche. Psychical content will be represented by a text whose essence is irreducibly graphic. The structure of the psychical apparatus will be represented by a writing machine. What questions will these representations impose upon us? We shall not have to ask if a writing apparatus—for example, the one described in the "Note on the Mystic Writing Pad"—is a good metaphor for representing the working of the psyche, but rather what apparatus we must create in order to represent psychical writing; and we shall have to ask what the imitation, projected and liberated in a machine, of something like psychical writing might mean. And not if the

psyche is indeed a kind of text, but: what is a text, and what must the psyche be if it can be represented by a text? For if there is neither machine nor text without psychical origin, there is no domain of the psychic without text. Finally, what must be the relationship between psyche, writing, and spacing for such a metaphoric transition to be possible, not only, nor primarily, within theoretical discourse, but within the history of psyche, text, and technology?[7]

Therefore, what is at stake in Derrida's reading of Freud consists in the possibility of extending the notions of "text" and "writing" to the functioning of the psychic system, while verifying the consequences of such an extension on these notions, that is, on their traditional definition, on what we understand as empirical writing and text. The result is a generalization of these notions that would be even more radical and, finally, would account for the conditions of possibility of life in general. The analysis starts from *Project for a Scientific Psychology* (1895), where, according to Derrida, the necessity of arche-writing emerges within a context that is analogous to the one examined in Husserl. Like retention for Husserl, memory for Freud is the irreducible condition of the constitution of consciousness: "Memory, thus, is not a psychical property among others; it is the *very* essence of the psyche: resistance, and precisely, thereby, an opening to the effraction of the trace."[8] However, as Derrida remarks, in the *Project*, the question was to explain memory in the wake of the natural sciences in order "to furnish a psychology that shall be a natural science: that is, to represent psychical processes as quantitatively determined states of specifiable material particles."[9] In particular, the *Project* is inscribed within the more general framework of the evolutionary theory of the time where "the struggle for survival" represents the determining factor of the evolution of life. Freud grounds the *Project* on the "principle of neuronal inertia," according to which the neurons constitutive of the nervous system tend to set themselves free from the quantity of excitation [ϱή] produced by internal and external stimuli. It is necessary to quote the "scientific" premises of the *Project* extensively because they allow us to understand in detail some passages from "Freud and the Scene of Writing" that appear enigmatic. For the moment, I observe that, according to Freud, once the nervous system reaches in the course of evolution a structure that is sufficiently complex, it must face two threats within the framework of the struggle for survival that determine its development with respect to the environment: 1. External stimuli against which the nervous system develops processes of immediate motor reaction that permit it to liberate itself immediately from the quantity of excitation through the reaction to the stimulus (*flight from the stimulus*):

> If we go further back from here, we can in the first instance link the nervous system, as inheritor of the general irritability of protoplasm, with the irritable external surface of an organism, which is interrupted by considerable stretches of non-irritable surface. A primary nervous system makes use of this ϱή which it has thus acquired, by giving it off through a connecting path to the muscular mechanisms, and in that way keeps itself free from stimulus. This discharge represents the primary function of the nervous system. Here is room for the development of a secondary function. For among the paths of discharge those are preferred and retained which involve a cessation of the stimulus: *flight from the stimulus*. Here in general there is a proportion between the ϱ of excitation and the effort necessary for the flight from stimulus, so that the principle of *inertia* is not upset by this.[10]

2. Internal stimuli (hunger, respiration, and sexuality, which are commonly considered the first determinations of the concept of *drive* [*Trieb*] elaborated by Freud): given the contingent environmental conditions on which the possibility of reaction hinges, the nervous system develops against them a reserve of the quantity of excitation derived from internal as well as external stimuli, which allows for the reinvestment or discharge of the stimuli in a deferred time, through a specific reaction:

> The principle of inertia is, however, broken through from the first owing to another circumstance. With an increasing complexity of the interior of the organism, the nervous system receives stimuli from the somatic element itself—endogenous stimuli—which have equally to be discharged. These have their origin in the cells of the body and give rise to the major needs: hunger, respiration, sexuality. From these the organism cannot withdraw as it does from external stimuli; it cannot employ their ϱ for flight from the stimulus. They only cease subject to particular conditions, which must be realized in the external world. For instance, the need for nourishment. In order to accomplish such an action (which deserves to be named "specific"), an effort is required

which is independent of endogenous Qή and in general greater, since the individual is being subjected to conditions which may be described as *the exigencies of life*. In consequence, the nervous system is obliged to abandon its original trend to inertia (that is, to bringing the level of Qή to zero) it must put up with maintaining a store [*Vorrat*] of Qή sufficient to meet the demand for a specific action. Nevertheless, the manner in which it does this shows that the same trend persist, modified into an endeavor at least to keep the Qή as low as possible and to guard against any increase of it. All the functions of the nervous system can be comprised either under the aspect of the primary function or of the secondary one imposed by the exigencies of life.[11]

It is within this scientific framework, which is manifestly evolutionistic, that Freud advances the hypothesis of the contact-barriers that, inserted among the neurons and thus resistant to internal and external stimuli, make the conservation of the quantity of excitation in the psychic system possible.[12] For Freud, only this hypothesis would allow us to explain the genesis and structure of memory as the response of the living to the urgencies dictated by life and thus by the struggle for survival: "Furthermore, the theory of contact-barriers can be turned to advantage as follows. A main characteristic of nervous tissue is memory: that is, quite generally, a capacity for being permanently altered by single occurrences."[13] In particular, through the hypothesis of the contact-barriers, Freud overcomes a difficulty that no explanation of memory can elude: on the one hand, the psychic system conserves the impressions received from experience through perception; on the other, perception is always available to receive new impressions. Freud faces this problem by distinguishing two classes of neurons:

There are two classes of neurones: 1. Those which allow Qή to pass through as though they had no contact-barriers and which, accordingly, after each passage of excitation are in the same state as before, and 2. Those whose contact-barriers make themselves felt, so that they only allow Qή to pass through with difficulty or partially. The latter class may, after each excitation, be in a different state from before and they thus afford a *possibility of representing memory*. Thus there are *permeable* neurones (offering no resistance and retaining nothing), which serve for perception, and *impermeable* ones (loaded with resistance, and holding back ϱή), which are the vehicles of memory and so probably of psychical processes in general. Henceforward I shall call the former system of neurones φ and the latter ψ.[14]

In this passage Freud brings out the notion of *"Bahnung"* (translated as *"frayage"* in French and "breaching" in English), which becomes essential for Derrida.[15] It allows Derrida to think the genesis of the retentional trace as the resistance offered to the impression and thus the trace as the effect of a contact between two forces and not of the impression as such; but, above all, it explains why memory conserves one impression rather than another on the basis of the differences of force among impressions:

> An equality of resistance to breaching, or an equivalence of the breaching forces, would eliminate any preference in the choice of itinerary. Memory would be paralyzed. It is the difference between breaches which is the true origin of memory, and thus of the psyche. Only this difference enables a "pathway to be preferred [*Wegbevorzugung*]." "Memory is represented [*dargestellt*] by the differences in the facilitations of the ψ-neurones" (*SPS*, 300). We then must not say that breaching without difference is insufficient for memory; it must be stipulated that there is no pure breaching without difference. Trace as memory is not a pure breaching that might be reappropriated at any time as simple presence; it is rather the ungraspable and invisible difference between breaches. We thus already know that psychic life is neither the transparency of meaning nor the opacity of force but the difference within the exertion of forces.[16]

Through Freud's description of the biological genesis of the psychic system and thus of the retentional trace, we approach the overcoming of phenomenology towards arche-writing according to the coordinates advanced in *Edmund Husserl's "Origin of Geometry"* and confirmed in *Speech and Phenomena*. As the retentional trace is the effect of the encounter between the force of impression and the resistance of the surface of impression, we cannot go back to the trace retained in the memory in the living and punctual presence of the impression that produced the trace itself. The latter can only refer to the impression through the *Bahnung* inscribed in the psychic system and not to the supposedly full presence, to the supposedly living present of the impression. Therefore, the retentional trace is already a differential trace. For Derrida, in the wake of Freud, not only can we identify iterability with the irreducible condition of the constitution of the retentional trace and thus of memory. Freud claims that actually there are two factors that contribute to *Bahnung* and thus to the constitution of memory: "What, then, does the *facilitation* in the ψ neurons depend on? According to psychological knowledge, the memory of an experience (that is, its continuing operative power) depends on a factor which is called the magnitude of the impression and the frequency with which the same impression is repeated."[17] Derrida forces Freud's interpretation, arguing beyond him that the repetition of the impression is not merely a concomitant and even supplementary factor that, alongside the force of impression, contributes to the genesis of memory; rather, it is its irreducible condition of possibility. The possibility of repetition is the condition of *Bahnung* and thus of the inscription of the force of impression in the psychic system, that is, of the recognition of the other occurrences of the same impression. In order to receive other impressions as repetitions of the same impression, the trace of the supposedly first impression must be already constituted in view of its repetition:

> For repetition does not happen to an initial impression; its possibility is already there, in the resistance offered the first time by the psychical neurons. Resistance itself is possible only if the opposition of forces lasts and is repeated at the beginning. It is the very idea of a first time which becomes enigmatic . . . We may still maintain that in the *first time* of the contact between two forces, repetition has begun.[18]

It seems clear that Derrida finds in the *Project*, that is, in the description of memory elaborated by Freud on a biological basis, the conditions of possibility of retention that he had uncovered in *Edmund Husserl's "Origin of Geometry"*: iterability as the irreducible condition of the differential character of the retentional trace. These conditions that Husserl ascribed only to the written trace, to empirical writing, will impose to Freud, in the works that follow the *Project* up to a "A Note upon the Mystic Writing-Pad," the recourse to writing and to more and more complex devices of writing that permit him to describe the functioning of the psychic system and, for Derrida, the notion of arche-writing. It is not by chance that the term "arche-writing" appears only once in *Writing and Difference*, precisely in the note that closes "Freud and the Scene of Writing," a note added only at the moment of the publication in *Writing and Difference* in 1967.[19] Therefore, through the reading of the *Project*, Derrida comes to the description of the genesis and structure of retention that overcomes the limits of the Husserlian position, still anchored to the principle of presence. And this is possible along the way opened up in *Edmund Husserl's "Origin of Geometry,"* that is, by tracing the condition of possibility of retention before the constituted consciousness and thus by understanding the genesis of consciousness as a biological emergence, as the specific manifestation of a particular form of life. But Derrida does not limit himself to this step beyond phenomenology in view of arche-writing. In "Freud and the Scene of Writing," he inscribes the description of the biological genesis of the psychic system within a broader horizon that accounts for this genesis. Freud's description, based on the evolutionary principle of the struggle for survival, would be only a moment, a specific articulation, related to the constitution of the psychic system, of the more general dynamic that constitutes its irreducible condition of possibility, the dynamic of différance:

> All these differences in the production of the trace may be reinterpreted as moments of deferring [*différance*]. In accordance with a motif which will continue to dominate Freud's thinking, this movement is described as the effort of life to protect itself by *deferring* a dangerous cathexis, that is, by constituting a reserve (*Vorrat*). The threatening expenditure or presence are deferred with the help of breaching or repetition. Is this not already the detour (*Aufschub*, lit. delay) which institutes the relation of pleasure to reality (cf. *Beyond the Pleasure Principle*)? Is it not already death at the origin of a life which can defend itself against death only through an economy of death, through deferment, repetition, reserve?[20]

Arche-writing, as the condition of memory and thus of the psychic system, first, and of consciousness, then, is an articulation of the more general dynamic of difference. It is the effect that the process of retentional iteration—that affects everything that "is"—produces at the level of the constitution of the psychic system, as a response to the urgencies dictated by life and thus by the struggle for survival. Hence death—the accidental possibility of death as well as its structural necessity—is always already inscribed in the heart of the living as its irreducible condition of possibility, as the condition of the genesis and structure of the living itself that is organized and evolves in order to escape the possibility of accidental death determined by environmental conditions and to delay the possibility of the structural death that every form of life undergoes. Derrida draws another, even more radical conclusion: différance is the most general condition of possibility of life; it regulates the organization and evolution of the living in general:

> No doubt life protects itself by repetition, trace, différance (deferral). But we must be wary of this formulation: there is no life present *at first* which would then come to protect, postpone, or reserve itself in différance. The latter constitutes the essence of life. Or rather: as différance is not an essence, as it is not anything, *it is not* life, if Being is determined as *ousia*, presence, essence/existence, substance or subject. Life must be thought of as trace before Being may be determined as presence. This is the only condition on which we can say that life *is* death, that repetition and the beyond of the pleasure principle are native and congenital to that which they transgress.[21]

We can take this passage as the germ of biodeconstruction. Here the motives that will be developed and disseminated throughout Derrida's oeuvre, in particular in the seminar *La vie la mort*, in the unedited sessions as well as in the published ones, are gathered together: différance as the condition of possibility of life and the consequential necessity of rethinking the relationship between life and death as a differential/differing relationship. Although here the *trace* works as the structural element of the organization of the psychic system, whose weave is therefore textual, it already announces the extension and generalization of the textual paradigm to the living in general, which will be elaborated in the first part of the seminar. Bound to the theory of evolution and thus recognized as the condition not only of the structural organization but also of the behavior of the living, *survival* already alludes to the way the differential/differing relationship life/death conditions the human experience and its symbolical elaboration. The *economy of death* and, thus, the irreducible necessity of the exposure to death in the interest of life disclose the still distant horizon of autoimmunity. Furthermore, the references to *Beyond the Pleasure Principle*[22] in the quoted passage announce the path that Derrida will follow in the last part of the seminar *La vie la mort*, published as "To Speculate—On 'Freud'" in *The Post Card*.

However, before beginning to read the seminar, it is necessary to go through *Of Grammatology* and draw attention to the first stage of development of these coordinates and, thus, to the formalization of the program that Derrida will follow in the seminar.

The Arche-Writing of the History of Life

Of Grammatology begins by emphasizing a certain inflation of the use of the word "writing" across all fields of knowledge as well as the discourses of arts, politics, military strategy, and sport: "All this to describe not only the system of notation secondarily connected with these activities but the essence and the content of these activities themselves."[23] In particular, Derrida observes, the recourse to writing as an explanatory model produces remarkable effects in two specific fields. These effects are so remarkable that we can find there the signs of a deconstruction to come or, rather, already at work:

> It is also in this sense that the contemporary biologist speaks of writing and *pro-gram* in relation to the most elementary processes of information within the living cell. And, finally, whether it has essential limits or not, the entire field covered by the cybernetic *program* will be the field of writing. If the theory of cybernetics is by itself to oust all metaphysical concepts—including the concepts of soul, of life, of value, of choice, of memory—which until recently served to separate the machine from man, it must conserve the notion of writing, trace, *grammè*, or grapheme, until its own historico-metaphysical character is also exposed. Even before being determined as human (with all the distinctive characteristics that have always been attributed to man and the entire system of significations that they imply) or nonhuman, the *grammè*—or the grapheme—would thus name the element. An element without simplicity. An element, whether it is understood as the medium or as the irreducible atom, of the arche-synthesis in general, of what one must forbid oneself to define within the system of oppositions of metaphysics, of what consequently one should not even call *experience* in general, that is to say the origin of *meaning* in general.[24]

Biology and cybernetics are the sciences in which the recourse to the notions of writing, trace, and, in particular, *grammè* caused the most radical solicitations in the system of hierarchized oppositions that structures Western thought, the so-called metaphysics of presence. Derrida refers to Norbert Wiener, one of the founding fathers of cybernetics, who is mentioned in a related footnote. However, for Derrida, Wiener is still a prisoner of metaphysics, as his naive use of the notions imported from the field of life suggests: "Wiener, for example, while abandoning 'semantics,' and the opposition, judged by him as too crude and too general, between animate and inanimate etc., nevertheless continues to use expressions like 'organs of sense,' 'motor organs,' etc. to qualify the parts of the machine."[25] To grasp the radical bearing of the *grammè*, we should rather look at what happens in the field of biology, where the notions introduced by cybernetics make it possible to understand and account for the most elementary processes of the organization of the living. From this perspective, we could even find in the structure of arche-synthesis (what Derrida later calls différance) the structural condition of the basic processes of life and, thus, the origin of *sense* in general.[26] Therefore, drawing together biology and cybernetics is not at all accidental. Evidently Derrida was aware of the revolution that the contribution of cybernetics was provoking in the field of genetics and, thus, in the life sciences in general.[27] In particular, he seems to be aware of the work of Jacques Monod, François Jacob, and André Lwoff, who were awarded the Nobel Prize for medicine precisely in 1965.[28] Derrida had likely read the proceedings of the conference held in Royaumont in 1962, *Le concept d'information dans la science contemporaine* (*The concept of information in contemporary sciences*), which was attended not only by Wiener and Moles, the fathers of cybernetics, but also by Lwoff and Jean Hyppolite, Derrida's mentor.[29] In his paper, "Le concept d'information dans la biologie moléculaire," Lwoff explicitly recurs to writing and text as metaphors to account for the genesis and structure of the living.[30] It would be of great interest to document Derrida's awareness of those proceedings because, as we will see, in the first part of the seminar *La vie la mort*, which is essentially devoted to the reading of François Jacob's *The Logic of Life*, Derrida again takes up those metaphors with effects that are both surprising and of incalculable relevance. Furthermore, given his close relationship with Georges Canguilhem, we could suppose that, at the moment of the first version of *Of Grammatology*, Derrida had already known about the essay "Le Concept et la vie," published by Canguilhem in 1966 and dedicated to the interpretation of the relationship between biology and cybernetics within the framework of the Aristotelian and Hegelian philosophy of life. There we can find a discussion of the recourse to a scriptural metaphor that would describe the structure and function of the genetic program.[31] However, the linguistic-scriptural metaphor, imported by cybernetics, was already diffused in genetic and molecular biology before it was adopted by Monod, Lwoff, and Jacob to account for the combinatorial structure of DNA and RNA and their relation of translation, although Jacob's role in the heuristic stabilization of the metaphor has been widely acknowledged.[32] However, in *Of Grammatology*, Derrida does not limit himself to a generic reference to the articulation of cybernetics and life sciences. They enter the stage at the very moment of the genesis of the *grammè* and, more precisely, of the genesis of arche-writing. In this case, the reference is explicit, namely, André Leroi-Gourhan's *Gesture and Speech* (1964), one of the "three relevant publications" that had been for Derrida the occasion for the two essays published in *Critique* (December 1965 and January 1966), which were later developed into *Of Grammatology*.[33] In the two volumes of *Gesture and Speech*, Leroi-Gourhan reconstructs the adventure of man from an evolutionist perspective based on the results of paleontology and, thus, begins from the zoological, that is, the anatomic and neurophysiological, structure of man and of his oldest ancestors. From this perspective, *Homo sapiens* would be the outcome of the evolution that led these ancestors to assume the erect posture that would have progressively freed the cranial vault, which is, in turn, the cause of the specific development of the human brain. The complexity of the human brain would be in fact the only specific trait that demarcates us from the other higher animals (even if it is a difference in degree and not in essence). In particular, Leroi-Gourhan identifies the first proofs of the emergence of *Homo sapiens* with the first traces of writing [*graphie*], which date back to 30,000 years before our era and "denote a deliberate repetition."[34] They are regular engravings deprived of any evident symbolical reference, in which it is possible to recognize the functional structure of iterability that characterizes also the forms of conventional writing that will appear only 20,000 years later. Therefore, the structural possibility of writing in its restricted sense had been liberated since the dawn of *Homo sapiens*. All this is evidently decisive for Derrida, for various reasons I have discussed elsewhere.[35] Here I remark that it is precisely in this context that the reference to Leroi-Gourhan in a key passage of *Of Grammatology* is inscribed:

Leroi-Gourhan no longer describes the unity of man and the human adventure thus by the simple possibility of the *graphie* in general; rather as a stage or an articulation in the history of life—of what I have called différance—as the history of the *grammè*. Instead of having recourse to the concepts that habitually serve to distinguish man from other living beings (instinct and intelligence, absence or presence of speech, of society, of economy, etc. etc.), the notion of *program* is invoked. It must of course be understood in the cybernetic sense, but cybernetics is itself intelligible only in terms of a history of the possibilities of the trace as the unity of a double movement of protention and retention. This movement goes far beyond the possibilities of the "intentional consciousness." It is an emergence that makes the *grammè* appear *as such* (that is to say according to a new structure of nonpresence) and undoubtedly makes possible the emergence of the systems of writing in the narrow sense. Since "genetic inscription" and the "short programmatic chains" regulating the behavior of the amoeba or the annelid up to the passage beyond alphabetic writing to the orders of the logos and of a certain *homo sapiens*, the possibility of the *grammè* structures the movement of its history according to rigorously original levels, types, and rhythms. But one cannot think them without the most general concept of the *grammè*. That is irreducible and ungraspable. If the expression ventured by Leroi-Gourhan is accepted, one could speak of a "liberation of memory," of an exteriorization of the trace, that has always already begun and, yet, becomes larger and larger [*d'une extériorisation toujours déjà commencée mais toujours plus grande de la trace*], which, beginning from the elementary programs of so-called "instinctive" behavior up to the constitution of electronic card-indexes and reading machines, enlarges différance and the possibility of putting in reserve: it at once and in the same movement constitutes and effaces so-called conscious subjectivity, its logos, and its theological attributes.[36]

Derrida does not simply say that, thanks to Leroi-Gourhan, it is possible to retrace the emergence of writing, in its restricted sense, back to a much greater history, which would find its roots in the emergence of

Homo sapiens in prehistory. Rather, the notion of *grammè* makes it possible to inscribe this history in an even greater one: the history of life itself, namely, the evolution of life in general, which is governed by the law of survival. Therefore, the *grammè* would allow us to understand the most general structure of life and of its evolution, of which writing in its restricted sense would be only a moment. Above all, and this is what interests us more, the *grammè* would allow us to point out that *différance* is a genetico-structural condition of the life of the living and of its evolution. From this perspective, we should first keep in mind the context of Derrida's reference to Leroi-Gourhan. In a footnote, Derrida provides some bibliographical references to *Gesture and Speech*, in particular, to Part II, which opens the second volume, "Memory and Rhythms."[37] In chapter I, titled "The Freeing of Memory," Leroi-Gourhan deconstructs the traditional philosophical opposition between humans and animals by depriving of any foundation the opposition of instinct and intelligence. Taking into account the organic and neuro-physiological structures that preside over behavior, there is no opposition between humans and animals but only a difference of degree. This difference measures the degree of the complexity of the nervous system and brain of the animal, including humans. At this point cybernetics enters the stage: by recurring explicitly to the notion of program elaborated in the field of cybernetics, Leroi-Gourhan compares the functioning of the nervous system and the brain of the animal to that of machines:

> The distinction between instinct and intelligence is of practical interest only at the extremes of the scale—in insects as well as in humans—and even there its real value is difficult to measure. The action programs of the lower vertebrates are closely conditioned by their internal environment and by external stimuli. The active behavior of an amoeba or an annelid can be reduced to short sequences triggered or prolonged by causes unrelated to what might be termed "automatic intelligence" as opposed to "intelligence based on reflection." Therefore it is not possible to trace the supposed transformation of instinct into intelligence by starting at the bottom end of creation and proceeding to the higher animals. The only fact that emerges from experimental study of animal behavior is the plasticity of an individual animal's behavior in relation to its specific means. This must be interpreted as a liberation, not from instinct, but from the fixed sequences established at the confluence of the

individual's internal biological environment and the exterior. The instinct-question is thus one of nervous apparatus rather than of the existence of a property peculiar to the animal condition. More precisely, the nervous system is not an instinct-producing machine but one that responds to internal and external demands by designing programs.[38]

In cybernetics, a *program* is an ordered sequence of inscribed instructions in a machine that makes it possible to respond to incoming information with outbound effects. For the animal, this program is the whole of the hereditary genetic instructions inscribed in the nervous system, which regulate the animal's behavior in the environment according to evolutionary laws of survival—that is, according to the necessity of avoiding danger and finding food and partners for reproduction. The difference in degree consists in the greater or lesser flexibility of the program, in its greater or smaller opening to possible variations in response—that is, in the ability to integrate the possible choices dictated by the influence of the environment and the group on the individual. But these possibilities are ultimately inscribed in the structure of the program. In the annelid and the amoeba, mentioned by Leroi-Gourhan[39] and recalled by Derrida, the program and its execution are very restricted because of the extreme simplicity of the nervous system. In humans, the program is very open because of the great complexity of the nervous system and the brain, which is able to operate a much greater number of connections than the brain of the other animals (an ability which is not at all exceptional because it depends on the liberation of the frontal vault of the cranium, due, in turn, to the assumption of the erect posture, which can be explained with some conditions dictated by the evolutionary law of survival):

> The characteristic feature of the individual behavior of mammals, at least so far as survival behavior is concerned, is the possibility of choice between action sequences, of checking the adequacy of each potential response to a given situation—a margin of control that varies from one species to another but is already very considerable in carnivores and primates. If we pursued the analogy with electronic devices, we should have to add to the apparatus for triggering responses and memories another mechanism capable of comparing and of orienting the device toward a particular response. Within the sweep of

evolution, nervous systems in fact appear to have progressed in two opposite directions, some (those of insects and birds) toward behavior channeled more and more narrowly by the nervous apparatus and others (those of mammals and humans) toward a prodigious enrichment of the nerve pathways by connective elements capable of establishing connections between new situations and already experienced ones. The individual's memory, formed in the earliest period of life, then takes precedence over the species memory, which is merely the result of the hereditary arrangement of the nervous system.[40]

Now, going back to the passage in which Derrida refers to Leroi-Gourhan, we can argue that, for Derrida, the description of the evolutionary structure of the behavior of the animal, as it is elaborated by Leroi-Gourhan through the mode of the cybernetic program, necessarily implies the structure of arche-writing as its irreducible condition and thus allows him to acknowledge the genesis of arche-writing from the animal and, thus, much further from the constituted intentional consciousness. If the life of the animal depends on interaction with the environment as it is described by Leroi-Gourhan, then it is regulated by the possibility of elaborating iterable traces: before any opposition between humans and animals, the animal in general must be endowed with a structure of retention and protention and, thus, must be capable of memory. In the wake of Leroi-Gourhan, Derrida can say that the intentional consciousness described by phenomenology is only a particular emergence of the general structure of the living programmed to respond to the necessity of survival. Therefore, arche-writing would be already at stake in the animal life, in the necessity of recognizing sources of food, reproductive partners, and dangers. Leroi-Gourhan had already led us to these conclusions: in *Gesture and Speech*, he proposes an extension of the notion of memory that seems to anticipate, almost literally, the notion of arche-writing:

> In this book the term "memory" is used in a very broad sense. It is not a property of the intelligence, but, whatever its nature, it certainly serves as the medium for action sequences [*le support sur lequel s'inscrivent les chaînes d'actes*]. That being so, we can speak of a "species-related memory" in connection with the establishment of behavior patterns in animal species, of an "ethnic" memory that ensures the reproduction of behavior patterns

in human societies, and similarly of an "artificial" memory in its most recent form that—without referring to either instinct or thought—ensures the reproduction of sequences of mechanical actions.[41]

Apparently, memory as the support of inscription, as the cause and, at the same time, the effect of the iterable trace, is not an invention by Derrida. Anyway, it is possible to retrace the extension of the notion of arche-writing as the condition of possibility of memory back to the extension proposed by Leroi-Gourhan with regard to memory (from the perspective of its animal, pre-cultural genesis as much as of its exteriorization and technical evolution through the different devices of writing). In *Of Grammatology*, Derrida writes:

> If the trace, arche-phenomenon of "memory," which must be thought before the opposition of nature and culture, animality and humanity, etc., belongs to the very movement of signification, then signification is a priori written, whether inscribed or not, in one form or another, in a "sensible" and "spatial" element that is called "exterior." Arche-writing, at first the possibility of the spoken word, then of the "*graphie*" in the narrow sense, the birthplace of "usurpation," denounced from Plato to Saussure, this trace is the opening of the first exteriority in general, the enigmatic relationship of the living to its other and of an inside to an outside: spacing. The outside, "spatial" and "objective" exteriority which we believe we know as the most familiar thing in the world, as familiarity itself, would not appear without the grammè, without différance as temporalization, without the non-presense of the other inscribed within the sense of the present, without the relationship with death as the concrete structure of the living present.[42]

At this point, it is necessary to think arche-writing as the structural condition of possibility of the life of the living, whose genesis responds to the laws of evolution and, ultimately, to the law of survival in the environment. In particular, we must think arche-writing as the condition of possibility of the constitution of the nervous system of all forms of animal life and not only of the human being; a condition that in the course of evolution—of the history of life as différance—is articulated and specified through different degrees of complexity in organization and function, up to the stage of human consciousness and beyond, that is, up to its technical exteriorization in the devices of inscription and iteration of memory, from the first types of primitive graphism to the technology of information. It is from this perspective that Derrida reappropriates the central thesis of Leroi-Gourhan's work, according to which human evolution, once it has reached a certain level from a biological and physical point of view, goes on as a technical evolution.[43] The passage from *Of Grammatology* in which Derrida addresses Lévi-Strauss's thesis that writing would be born in order to subjugate the subaltern members of a community confirms what can already be grasped from the above-mentioned passage dedicated to Leroi-Gourhan:

> This phenomenon is produced from the very onset of sedentarization; with the constitution of stocks at the origin of agricultural societies. Here things are so patent that the empirical illustration that Levi-Strauss sketches could be infinitely enriched. This entire structure appears as soon as a society begins to live as a society, that is to say from the origin of life in general, when, at very heterogeneous levels of organization and complexity, it is possible to defer presence, that is to say expense or consumption, and to organize production, that is to say reserve in general. This is produced well before the appearance of writing in the narrow sense, but it is true, and one cannot ignore it, that the appearance of certain systems of writing three or four thousand years ago was an extraordinary leap in the history of life. All the more extraordinary because a prodigious expansion of the power of différance was not accompanied, at least during these millennia, by any notable transformation of the organism. It is precisely the property of the power of différance to modify life less and less as it spreads out more and more.[44]

The possibility of the reserve in general, namely, arche-writing, occurs much earlier than the beginning of sedentary agricultural societies and thus of empirical writing, but also earlier than the first forms of primitive graphism. For Derrida, it goes back to the origin of life itself. There he finds the manifestation of the most general law that regulates the organization of the life of the living and not only its behavior in the environment: the law of différance. It is worth recalling that, in the passage dedicated to Leroi-Gourhan, Derrida had already referred to that possibility:

Since "genetic inscription" and the "short programmatic chains" regulating the behavior of the amoeba or the annelid up to the passage beyond alphabetic writing to the orders of the logos and of a certain *homo sapiens*, the possibility of the *grammè* structures the movement of its history according to rigorously original levels, types, and rhythms.[45]

We can see that an interpretation of différance as the structural condition of the life of the living, from its elementary biological structure to the psychic system that presides over the relationship between the living and alterity in general, takes shape in *Of Grammatology*.[46] Therefore, even earlier than the psychic system, arche-writing structures the living itself as a text, as a weave of iterable traces, developed in view of its iteration and thus its survival; a differential/differing iteration that, in the case of the human being, is propagated across the cultural environment.[47] But to justify this perspective we shall turn to genetic biology. This urgency is made explicit, as we saw above, at the beginning of *Of Grammatology*, when Derrida referred to the recourse to writing and cybernetics in biology: "It is also in this sense that the contemporary biologist speaks of writing and *pro-gram* in relation to the most elementary processes of information within the living cell."[48] This is the trace that Derrida will follow in the first part of the seminar *La vie la mort*.

II

Between Life and Death: Différance

The seminar *La vie la mort*, which Derrida taught at the Ecole Normale in Paris in 1975, consists of fourteen sessions.[1] In the first session, Derrida introduces the seminar and begins to read François Jacob's *The Logic of Life* (1970). The second session is dedicated to Nietzsche, to the relation between the philosopher's life and his oeuvre as developed in his writings, in particular in *Ecce Homo*. In other words, what is at stake is the problem of auto-bio-thanato-graphy, which Derrida treats extensively in his published works and which we are not examining in this volume. I limit myself to observing only that this problem comes out of the intersection between life, as the object of discourse (mythical, religious, philosophical, scientific, psychoanalytic, literary, and so forth) and the life of the subject of the discourse itself. This session has been published in full, with a few interpolations and the addition of footnotes, in *Otobiographies*.[2] The first part of the third session also concerns Nietzsche, his theory of the physiological origin of the metaphor, and, in more detail, the recourse to phenomena borrowed from biological life as a metaphorical resource to describe the state of degeneration of German academic institutions in *On the Future of Our Educational Institutions*. The second part addresses the metaphor/concept opposition within scientific discourse and, in particular, in relation to Jacob's *The Logic of Life* and Canguilhem's "The Concept of Life." In Sessions 4 to 6, Derrida goes back to the analysis of *The Logic of Life*. Sessions 7 to 10 are dedicated to Nietzsche and the deconstruction of Heidegger's critique of Nietzsche's so-called "biologism." The complete text of Session 8 and the first three pages of Session 9 were published in "Interpreting Signatures (Nietzsche/Heidegger): Two Questions"[3] and address the questions of the proper name and signature and of the role these concepts play in Nietzsche's oeuvre. They are inscribed in the context of the critique of Heidegger's interpretation of Nietzsche as a metaphysical thinker and, more generally, of his "biologism."[4] This part is less extensive, at least if measured against Derrida's aim to examine Nietzsche's relationship with the life sciences and Heidegger's interpretation, which is, in any case, fiercely criticized. Unfortunately, this critique is merely sketched out, and so I shall not discuss it. The last four sessions are dedicated to a reading of Freud's *Beyond the Pleasure Principle* and consist of the palimpsest of "To Speculate—On Freud," published in *The Post Card* (1980). More precisely, the text published as "To Speculate—On Freud" is not a mere rewriting of the part of the seminar dedicated to Freud. Its composition is more complex: it consists of the text of the seminar, which is reproduced integrally, almost literally (save a few, sometimes noteworthy, changes) and of parts that were added later and that either serve as short clarifications of the topics already treated in the seminar or amount to extensive elaborations of topics that are just sketched out in *La vie la mort*—the most extensive one concerns the intersection between the theoretical speculation on life and the life of the author engaged in this speculation. In "To Speculate—On Freud," Derrida deepens this theme apropos Freud, who completed the draft of *Beyond the Pleasure Principle* a short while after the premature death of his daughter Sophie and includes in the text a biographical episode dissimulated through a neutral observation (the ludic activity of his grandson Ernst, Sophie's son). In chapter V, when discussing the part of the seminar devoted to Freud, I refer to the text "To Speculate—On Freud" and indicate, where necessary, the passages corresponding to the seminar and the specific differences with the published text.

The Deconstruction of Life Between Hegel and Jacob

Derrida inaugurates the seminar with an explanation of the title, that is, of the choice of leaving the relationship between life and death undetermined:

> What did I do in announcing this seminar under the title *La vie-la mort*, that is, replacing with a hyphen [*trait d'union*] or with a spacing without trait, with a marked silence, the *and* that usually posits death with life, as juxtaposed or more properly

opposed to life? The relation of juxtaposition or opposition, the
relation of positing, the logic of positing (whether dialectic or
non-dialectic), this will be perhaps what precisely comes into
question with life death. In removing the *and* I didn't mean to
suggest that life and death are not two, or that they are not each
the other's other, but that this alterity or difference is not of
the order of what philosophy calls opposition (*Entgegensetzung*),
double position of two facing each other, like in Hegel, for
instance, where the concept of position and the position of the
concept, auto-position and opposition form the motor schemas
of dialectics and of a dialectics that affirms itself essentially as
a very powerful thinking of life and death, of what we call the
relations of life and death and, above all, where opposition,
contradiction (whether dialectical or not) is the process of the
transition of an opposite into the other, of the identification
sublating the one into the other.[5]

The stakes of the seminar are evident: freeing the thought of life and
death, of their relationship, of the border that separates and holds them
together, from the legacy of tradition and thus, above all, from the logic
of the hierarchically oriented oppositions that, for Derrida, structures the
Western philosophical tradition as a metaphysics. In *Positions*, which is from
only a few years before the seminar, while describing the "general strategy of
deconstruction," Derrida affirms the necessity of overturning this hierarchy:

> What interested me then, that I am attempting to pursue along
> other lines now, was, at the same time as a "general economy,"
> a kind of *general strategy of deconstruction*. The latter is to avoid
> both simply *neutralizing* the binary oppositions of metaphysics
> and simply *residing* within the closed field of these oppositions,
> thereby confirming it. Therefore we must proceed using a double
> gesture, according to a unity that is both systematic and in and
> of itself divided, a double writing, that is, a writing that is in
> and of itself multiple, what I called, in *La double séance*, a *double
> science*. On the one hand, we must traverse a phase of *overturning*.
> To do justice to this necessity is to recognize that in a classical
> philosophical opposition we are not dealing with the peaceful
> coexistence of a *vis-a-vis*, but rather with a violent hierarchy.
> One of the two terms governs the other (axiologically, logically,

etc.), or has the upper hand. To deconstruct the opposition, first of all, is to overturn the hierarchy at a given moment. To overlook this phase of overturning is to forget the conflictual and subordinating structure of opposition.⁶

However, he also points out that deconstruction does not limit itself to the simple overturning of hierarchy, which would leave unaltered the field of intervention as well as the system of oppositions that structures that field. The effectiveness of deconstruction hinges on the introduction of a notion that destabilizes the metaphysical oppositions from within, in order to account for the differential/differing relationship as the irreducible condition of the constitution of the terms that are merely understood as opposed:

> To set to work, *within* the text of the history of philosophy, as well as *within* the so-called literary text certain marks, shall we say (I mentioned certain ones just now, there are many others), that *by analogy* (*I* underline) I have called undecidables, that is, unities of simulacrum, "false" verbal properties (nominal or semantic) that can no longer be included within philosophical (binary) opposition: but which, however, inhabit philosophical oppositions, resisting and disorganizing it, *without ever* constituting a third term, without ever leaving room for a solution in the form of speculative dialectics. (the *pharmakon* is neither remedy nor poison, neither good nor evil, neither the inside nor the outside, neither speech nor writing; the *supplement* is neither a plus nor a minus, neither an outside nor the complement of an inside, neither accident nor essence, . . . the *gram* is neither a signifier nor a signified, neither a sign nor a thing, neither presence nor an absence, neither a position nor a negation, etc. . . . Neither/nor: that is *simultaneously* either *or*; the mark is also the *marginal* limit, the *march*, etc.).⁷

For what concerns us here, this means that, on the one hand, the interest in death, which many interpreters assume as the dominant trait of Derrida's thought—think, for instance, of Roberto Esposito, who grounds on this assumption the alleged overcoming of deconstruction through biopolitics⁸—only consists in a preliminary moment (the overturning of the hierarchy). On the other hand, with "life death," Derrida prepares the ground for the introduction of a notion to which the deconstructive effects of his discourse are bound: "life death," life and death, neither life nor death, as long as they are considered by themselves and thus the one opposed to the other. Recalling the interview *Learning to Live Finally* that I quoted at the beginning of the book, I advance the following hypothesis: here Derrida lays the premises for the question of the survival, "*survie*" or "*survivance*," to which the bearing of the deconstruction of the traditional notion of life is linked, far beyond the seminar. Within these general coordinates, which orient deconstruction itself, in the opening scene of the seminar Derrida establishes in greater detail the textual coordinates that orient the attempt to elaborate a new thought of the relationship between life and death. The first coordinate, namely, Hegel, was already evident and, in any case, unavoidable:

> In insisting on the necessity of beginning with this kind of questions, with logical questions, if you like, in wondering whether the whole positional and oppositional logic in which the limit life/death has been and still is thought, not only is not powerful enough to think this limit but is itself produced as an effect of life death and thus must be reread as logic in general from this point of view, I'm pointing to two textual indicators that seem to me to impose themselves today. On the one hand, Hegel and above all Hegel's logic.[9]

Therefore, the seminar begins under the aegis of Hegel, whose authority consists in giving the most accomplished formulation of the philosophical concept of life. It is with Hegel that we must engage in order to formulate a new concept of life, a concept liberated from the metaphysical presuppositions that, to begin with the opposition between life and death, could silently influence, once again, the life sciences. This is what is at stake in contemporary biology, which aims to demarcate itself from philosophy, as well in the deconstruction of the concept of life. It is not by chance that François Jacob, who is considered an influential representative of this process of demarcation, stands for the second textual indicator:

> On the other hand, another textual indicator, that I will call the logic of the living, using the title of Jacob's book, today tends to decipher the living through the whole problematics of message, code, or even of genetic text (the living and not life, as these biologists remark, in order to distance themselves from what they

consider a bit too quickly the hypostatizing and substantialist compulsion of the philosopher, omitting, for instance, that Hegel demonstrates the necessity to go through the living, the living individual, as a necessary position within the syllogism of life, within the judgment, *Urteil*, of life that originarily splits itself to produce and re-produce itself), a modernity, then, that aims to decipher the living as a language (for the moment, I leave this term in all its indetermination) that depends on a logic. Between these two textual indicators lies the field in which we will be situated here.[10]

In *The Logic of Life*, Jacob rethinks the history through which biology came to understand the mechanisms of genetic heritage, to discover of the essential role that DNA plays in the production of the cell and thus the elementary unity of the life of the living, of all livings. Thanks to this discovery, biology could elaborate the logic that regulates the life of the living within the framework of the theory of evolution, namely, the logic of the reproduction or self-reproduction of the living, which, as Derrida remarks, Hegel had already considered the key for the understanding of the living. The theory of organized systems elaborated by cybernetics gave an essential contribution to the discovery of the logic of the living. Jacob also mentions Wiener and, like Leroi-Gourhan, recalls the analogy between machines and living forms,[11] but he goes further: it is possible to apply cybernetics not only to the behavior of animals, which are already organized with respect to the environment, but also to the mechanisms of genetic heredity and, thus, to the most elementary laws that regulate the formation of the living:[12]

> Organs, cells and molecules are thus united by a communication network. They constantly exchange signals and messages in the form of specific interactions between constituents. The flexibility of behavior depends on feedback loops; the rigidity of structures on the execution of a programme rigorously laid down. Heredity becomes the transfer of a message repeated from one generation to the next. The programmes of the structures to be produced are recorded in the nucleus of the egg . . . The plan of the organism is mapped out by a combinative system of chemical symbols. Heredity functions like the memory of a computer.[13]

The motives or discrete traits that play as the letters of the alphabet are the four elements of which the structure of DNA consists:

> It is a long polymer formed by the alignment of four sub-units, the four organic bases, repeated by millions and permuted along the chain, like the letters of the alphabet in a text. It is the order of these four sub-units that directs the order of the twenty sub-units in proteins. Everything then leads one to regard the sequence contained in genetic material as a series of instructions specifying molecular structures, and hence the properties of the cell; to consider the plan of an organism as a message transmitted from generation to generation.[14]

Therefore, on the one hand, Hegel is the representative of the philosophical determination of life, and on the other, Jacob is the representative of the life sciences and, in particular, of the emancipation of biology from philosophy, which would take place by grafting cybernetics onto the field of biology. On the one hand, through Hegel, Derrida deconstructs the metaphysical determination of life, grounded on the life/death opposition; on the other, through Jacob, he verifies the deconstructive effects of the recourse to the theory of information and in particular to the notions of "programme" and "writing" in the context of the life sciences. It is worth remarking, however, that, from the first pages, Derrida seems to be very suspicious about the alleged emancipation of biology from philosophy, at least for what concerns Jacob. It is precisely to the latter that Derrida refers between parentheses. Indeed, it is Jacob himself who justifies the choice of the title *The Logic of Life* (*La logique du vivant*) with the purpose of avoiding the hypostatization of life as an abstract, metaphysical entity, and thus it is for Jacob that Derrida recalls the analogous position of Hegel, who takes the living as a concrete point of departure for the reconstruction of the logic that would be proper of the living itself.[15] It seems that Derrida wishes to highlight a problem that has existed since the beginning: through the life/death opposition, the unconscious legacy of philosophy and metaphysics still affects modern biology, which however supposes to be liberated and thus risks neutralizing the innovations that the importation of cybernetics seems to enable. In the introduction to the seminar, Derrida argues that not only the Hegelian system but the entire philosophical tradition, of which the former aims to be the sublating accomplishment (*Aufhebung*), would be constructed in order to protect itself, that is, through the repression of another logic of the living, a logic according to which the relationship between life and death would no longer respond to the scheme of the oppositional logic of metaphysics by which life is being and death is non-being. Therefore, the syntagm "life death" points to another thought of life that would finally be able to account for another relationship between life and death, as well as for the reasons that led thought to repress this relationship, to hide it through the imposition of the opposition between life and death:

> By merely alluding to it at the outset, I wanted to mark three things: on the one hand, that the *and* of juxtaposition should not only be interrogated and thus suspended, while we wonder whether the relations of being and death really do come under what is called opposition or contradiction; but, more radically, if what is comprehended under the concept of position, opposition or juxtaposition, or even contradiction, is not constructed by a logic of "life death" which would be dissimulated—in view of which interest, this is the question—under a positional (oppositional, juxtapositional, dialectical) scheme, *as if* (I can only use the *as if* here, because I neither want nor can *oppose* a logic to the logic of opposition), as if the whole logic of opposition (logic of identity or dialectical logic, formal or dialectical logic), were a ruse, put forward by "life death" in order to dissimulate, guard, hide, host and forget—something. What? Something that, in any case, neither posits itself nor *is opposed* [*s'oppose*] any longer and that would no longer be something in this sense of position.[16]

The stakes of this seminar are therefore very high: the elaboration of another thought of life that, however, must first go through the deconstruction of the hierarchized opposition between life and death and thus through Hegel, namely, the speculative determination of the living.

Hegel: The Absolute Seed

From the opening of the seminar, Derrida draws attention to the double position of life in Hegelian philosophy, at the beginning and at the end, that is, in the *Philosophy of Nature*, in the transition from nature to subjective spirit, and in the first moment of the last syllogism of the *Science of Logic*, in which life is the first determination of the absolute Idea:

> In saying "life death" I do not mean to identify life and death, to say that life is death, pro-position that, as you know, can be recalled in multiple forms, through several, well known ways. The white trait between life and death does not come about to replace neither an *et* nor an *est*. In the dialectical logic of Hegel, the *est* of judgment comes here, as the place of the contradiction and of its *Aufhebung*, to enunciate that life is death, that it posits itself in its syllogism through the mediation of death, that, according to the dynamic and productive meaning of the word *est*, *est* is the process of death (the death of natural life *qua* the birth of spiritual life) through which the *est* becomes itself Life, the being of the *est* becomes again Life in a dissymmetry that I have attempted to analyze elsewhere and in which life is marked twice, as a moment of the process of the Idea and of being and, then, *without* death, which remains always natural, at the moment of the absolute Idea, at the end of the *great logic*, when Hegel writes "only the absolute idea is being, imperishable life, the truth that knows itself and is entirely true." At this moment, the last moment, life has no longer opposition, opposite, the opposition took place in it, in order for it to reappropriate itself, but life has no longer an other before itself. The *est* of life death is *of life, being is life*, death cannot be thought at all. Here it is where the oppositional logic leads us, when it gives the greatest attention to death (it is the case of Hegel): to the suppression of the opposition, to the sublation in the elevation of one term and in the process of its reappropriation. Life is this *reappropriation* of being, it is being: only the absolute Idea is being, imperishable life (non-death). Between the opposition (*et*) and the copulatory identification (*est*) there is no opposition, the opposition is the process of identification or reappropriation of being as life or of life as being.[17]

I remark that in this passage, death appears in the Hegelian system only as opposed to life, as the other of life, in order to be sublated—*Aufgehebt*—in the infinite and imperishable life of the absolute spirit. Death, qua opposed to life, is only thought in view of life; it is a moment of life that for Hegel must be well determined at the level of natural life and that would allow for the transition to the imperishable life of the absolute spirit. From this perspective, if within the system death is always thought in view of life and of a determination of life as absolute life, a life without death, we may easily suspect that the determination of natural and biological life also is affected by that determination. Hence for Derrida in the Hegelian system there is no death. Death would be the unthought-of, with the unavoidable consequence that there would be no thinking of life, of life as irreducibly affected by death, of natural and biological life, insofar as, within the system, life is determined in order to render the thought of the Idea, of the absolute Spirit qua imperishable life, possible. Therefore, the whole question depends on verifying which determination of natural and biological life authorizes Hegel to determine the absolute Idea and thus absolute spirit as life, but as life without death:

> If you follow the great syllogism of life at the end of Hegel's *Greater Logic* you will see that life, which is essentially a position (*Setzung*), a position of the Idea that posits itself through its three oppositions, namely, the living individual, the process of life and the species (*Gattung*), reappropriates itself as life through the opposition of death and is born as the life of the spirit in natural death, according to a movement that is marked everywhere in Hegel (I shall call it the movement of the phoenix) and that we must naturally come back to.[18]

Here the discussion of Hegel in the seminar stops. Derrida postpones his analysis to another undetermined place. It is not difficult to identify this place with *Glas* (1974), where he deals with the question of life in the Hegelian system precisely from the perspective suggested in the seminar:

> The Idea, immediate and natural life, relieves, abolishes and preserves, itself, dies in raising itself to the spiritual life. So life develops itself in contradiction and negativity; the metaphor between the two lives is only this movement of relieving negativity . . . The same movement in the *Encyclopedia*, at the end, concerning *Sa* [absolute Spirit]. The third term returning to immediacy, this return to simplicity being brought about by the relief [*relève: Aufhebung*] of difference and mediation, natural life

occupies both the end and the beginning. In their ontological sense, the metaphors are always of life; they put rhythm into the imperturbable equality of life, of being, of truth, of filiation: *physis*. Thus the Hegelian system commands that it be read as a book of life.[19]

And it is to *Glas* that Derrida alludes when, in the seminar, he calls "phoenicious movement" the dialectical process through which life sublates its merely natural determination, which ends with the biological death, in order to resurge as Spirit. Hegel often recurs to the phoenix that resurges from its ashes as the figure of that dialectical transition, of the movement of *Aufhebung* through which life appropriates itself by reducing death to a simple moment in view of life as the first determination of the Absolute Idea. In particular, at the end of the *Philosophy of Nature*, which announces the transition to the *Philosophy of Spirit*:

> This is the *transition from natural being into spirit*; nature has found its consummation in living being, and has made its peace by shifting into a higher sphere. Spirit has therefore issued forth from nature. The purpose of nature is to extinguish itself, and to break through its rind of immediate and sensuous being, to consume itself like a Phoenix in order to emerge from this externality rejuvenated as spirit. Nature has become distinct from itself in order to recognize itself again as Idea, and to reconcile itself with itself. To regard spirit thus, as having come forth from implicitness, and as having *become* a mere being-for-self, is however a onesided view. Nature is certainly that which is immediate, but as that which is distinct from spirit, it is nevertheless merely a relativity. As the negative of spirit, it is therefore merely a posited being. It is the power of free spirit which sublates this negativity; spirit is nature's antecedent and to an equal extent its consequent, it is not merely the metaphysical Idea of it. It is precisely because spirit constitutes the end of nature, that it is *antecedent* to it.[20]

In *Glas*, Derrida emphasizes several times Hegel's recourse to the phoenix in order to illustrate the speculative transition of life from nature to spirit and, as we shall see, dedicates a careful analysis to the aforementioned passage.[21] Before following the reference to *Glas*, in order to find an answer to this abyssal question, and, above all, to understand how, within the Hegelian system, death is neutralized or repressed by being posited in opposition to life and how this occurs in the transition from natural life to absolute spirit, it is necessary to step back and acknowledge that Derrida inherits from Bataille the idea, formulated against the Kojèvian evidence,[22] that death has never been present in the system and thus remains the unthought-of or the blind spot of Hegelian philosophy. In "From Restricted to General Economy," his essay on Bataille written in 1965, when discussing the struggle for recognition—*Der Kampf um Anerkennung*—between the master and the slave, Derrida remarks that Bataille demarcates himself from Kojève by proposing a displacement of tone in the reading of the Hegelian expression *Daransetzen des Lebens*, putting life in play. Bataille would be the first to point out that putting life in play is a play, a show, a mise-en-scène also in the theatrical meaning of the expression (one plays death), while life is not effectively put in play, risked. One puts life at stake precisely to avoid risking it seriously. According to Bataille, in order to find the profound motivations of this play, we should address the experience of sacrifice and thus find in the representation of death the sleight of hand that allows for recoiling before death qua an absolute loss without reserve. I quote a long passage from Derrida's reading, which discusses the implications that this perspective entails for the interpretation of the struggle for recognition, but also for the general economy of the system. Let me remark also that here we can find the first occurrence of the problem of survival in Derrida's oeuvre:

> Hegel clearly had proclaimed the necessity of the master's retaining the life that he exposes to risk. Without this economy of life, the "trial by death, however, cancels both the truth which was to result from it, and therewith the certainty of self altogether." To rush headlong into death pure and simple is thus to risk the absolute loss of meaning, in the extent to which meaning necessarily traverses the truth of the master and of self-consciousness. One risks losing the *effect* and profit of meaning which were the very stakes one hoped to win. Hegel called this mute and non-productive death, this death pure and simple, abstract negativity, in opposition to "the negation characteristic of consciousness, which cancels in such a way that it preserves and maintains what is sublated (*Die Negation des Bewusstseins welches so aufhebt, dass es das Aufgehobene aufbewahrt and erhält*), and thereby survives its being sublated (*und hiemit sein Aufgehoben-werden überlebt*).

In this experience self-consciousness becomes aware that life is as essential to it as pure self-consciousness." Burst of laughter from Bataille. Through a ruse of life, that is, of reason, life has thus stayed alive. Another concept of life had been surreptitiously put in its place, to remain there, never to be exceeded, any more than reason is ever exceeded. This life is not natural life, the biological existence put at stake in lordship, but an essential life that is welded to the first one, holding it back, making it work for the constitution of self-consciousness, truth, and meaning. Such is the truth of life. Through this recourse to the *Aufhebung*, which conserves the stakes, remains in control of the play, limiting it and elaborating it by giving it form and meaning, this economy of life restricts itself to conservation, to circulation and self-reproduction as the reproduction of meaning; henceforth, everything covered by the name lordship collapses into comedy.[23]

Here Derrida finds in the question of life the stakes of the repression of death at work in the Hegelian system, a repression of death that is at the same time a repression of life, of natural and biological life, which would take place through the surreptitious introduction of another concept of life replacing the former. The logic of this other life, which is not the natural and biological life, is the reproduction and conservation of the self. It is the same law that the biologist Jacob enunciates as the principle of the logic of the living. Therefore, we can understand why the seminar *La vie la mort* begins with Hegel. Biology, whose decisive progresses Derrida recognizes, must pay attention to its secret philosophical heritage that would bring it back to its metaphysical past. In the wake of Bataille, Derrida had already identified the blind spot of Hegelianism (and, thus, the distance between Bataille and Kojève):

The blind spot of Hegelianism, *around* which can be organized the representation of meaning, is the *point* at which destruction, suppression, death and sacrifice constitute so irreversible an expenditure, so radical a negativity—here we would have to say an expenditure and a negativity *without reserve*—*that* they can no longer be determined as negativity in a process or a system. In discourse (the unity of process and system), negativity is always the underside and accomplice of positivity . . . For negativity is a resource. In naming the without-reserve of absolute expenditure "abstract negativity," Hegel, through precipitation, blinded himself to that which he had laid bare under the rubric of negativity. And did so through precipitation toward the seriousness of meaning and the security of knowledge.[24]

We can go back to *Glas*, in which Derrida merely develops Bataille's insight—the repression of death qua absolute loss and its submission to *Aufhebung*—by finding it at work throughout the system. In particular, this is evident in those passages dedicated to the struggle for recognition in which Derrida brings the economical lexicon already adopted in Bataille's essay at its limits:

> This putting (in play, at pawn) must, as every investment, amortize itself and produce a profit; it works at my recognition by/through the other, at the posit(ion)ing of my living consciousness, my living freedom, my living mastery. Now death being in the programme, since I must *actually* risk it, I can always lose the profit of the operation: if I die, but just as well if I live. Life cannot stay in the incessant imminence of death. So I lose every time, with every blow, with every throw [*à tous les coup*]. The supreme contradiction that Hegel marks with less circumspection than he will in the *Phenomenology*.[25]

Derrida analyzes the struggle for recognition by referring to the Jena lectures in the *Philosophy of Spirit* because it is in this text that we can grasp the whole dialectical process in which the struggle is inscribed. The two consciousnesses must accept that, in order to be recognized, they renounce their being absolute, singular consciousnesses in favor of a third term, that is, of people, of the community that becomes the State, and, thus, they avoid struggle and death, and do not even risk it. Derrida never refers to Kojève in these passages from *Glas*, but a demystification of the latter's reading (as holding on to the opposition between the master and the slave) is evidently at work. Let me quote the conclusion of Derrida's analysis from *Glas*: "From that moment on, death, suicide, loss, through the passage to the people-spirit as absolute spirit, amortize themselves every time, with every blow, with every *coup*, in the political: at the end of operation, the absolute spirit records a profit in any case, death included [*la mort comprise*]."[26] It is time to recall what for Derrida is at play in this determination of death qua opposed to life, which allows the sublation, *Aufhebung*, of death itself in the absolute life of the absolute Spirit. Given that in the *Science of Logic* natural life is the first and immediate determination of the absolute Idea, which determination of natural and biological life permits the dialectical transition to the determination of the absolute Idea as life, life without death? Which hidden interest orients *Aufhebung* or the repression of the biological death, death as absolute loss? We must start by recognizing the systematic necessity of what in the *Logic* would appear as a simply illustrative metaphor: the germ, the seed as a botanic metaphor that helps us conceive of the absolute Idea as a living form:

> To this extent, it is the individuality of life itself, no longer *generated* out of its concept but out of the *actual* idea. At first, it is itself only the concept that still has to objectify itself, but a *concept which is actual—the germ of a living individual*. To *ordinary perception* what the concept is, and that the *subjective concept* has external actuality, are visibly present in it. For the germ of the living being is the complete concretion of individuality: it is where all the living being's diverse sides, its properties and articulated differences, are contained in their *entire determinateness*; where the at first *immaterial*, subjective totality is present undeveloped, simple and non-sensuous. Thus the germ is the whole living being in the inner form of the concept.[27]

Derrida remarks that, throughout his life and system, Hegel recurs to examples drawn from botany in order to describe the genesis and structure of the spirit and to determine it as life: germ, seed, tree, plant, from *The Spirit of Christianity* to the *Science of Logic*, passing through the *Encyclopedia* and the *Philosophy of Nature*. This cannot be explained either as an accident or as an illustrative metaphor. To understand the recourse to botanic metaphors we should take into account the role that the vegetal organism plays in the *Philosophy of Nature*, as the second moment of the last section, the "Organic Physics":

> § 337. The real nature of the body's totality constitutes the infinite process in which individuality determines itself as the particularity or finitude which it also negates, and returns into itself by reestablishing itself at the end of the process as the beginning. Consequently, this totality is an elevation into the

primary ideality of nature. It is however an *impregnated* and negative unity, which by relating itself to itself, has become essentially *selfcentred* and *subjective*. It is in this way that the Idea has reached the initial immediacy of *life*. Primarily, life is *shape*, or the universal type of life constituted by the *geological* organism. Secondly, it is the particular formal subjectivity of the *vegetable* organism. Thirdly, it is the individual and concrete subjectivity of the *animal* organism. The Idea has truth and actuality only in so far as it has subjectivity implicit within it (§ 215). As the mere immediacy of the Idea, life is thus external to itself, and is not life, but merely the corpse of the living process. It is the organism as the totality of the inanimate existence of mechanical and physical nature. Subjective animation begins with the vegetable organism, which is alive and therefore distinct from this inanimate existence. The parts of the individual plant are themselves individuals however, so that the relations between them are still exterior. The animal organism is so developed however, that the differences of its formation only have an essential existence as its members, whereby they constitute its subjectivity.[28]

The first determination of life is nature itself, the geological organism, as the totality of the physical world, and, however, this is only an immediate and abstract determination, insofar as it encompasses nature as the whole in which there is yet no distinction between the living and the non-living. Hence, the vegetal organism is the first determination of biological life as such and, thus, the first moment of natural life as well as the first moment of the Idea: the plant, indeed, as the living individual, is the first manifestation of the dialectical structure of subjectivity in nature. From this perspective, the absolute life of the spirit represents the accomplishment of a dialectical process that finds in the life of the vegetal organism its first moment and in the life of the animal organism, and, thus, of man, the middle term. Therefore, at the end of this process of sublation or *Aufhebung*, the absolute life of the spirit is accomplished according to the formal, subjective structure of the vegetal organism, developed in its dialectical content according to the determination of the life of the animal organism that the life of the Spirit retains in itself as the middle term.

It is worth asking on what this distribution of life in the *Philosophy of Nature* is grounded: why is the vegetal organism found in this privileged position from the point of view of the formal structure? Why does the animal organism only consist in a concrete development of the "differences of its formation," and why will it be sublated in the life of the spirit, which is achieved (in itself and for itself) according to the structure anticipated in the still-immediate life of the vegetal organism (in itself) and through the mediation of the animal life (the being for an other)? Is the reason of this distribution wholly inherent in the dialectic of nature or does it respond to a more general speculative interest? Derrida analyzes a text whose title would be *The Determination of the Spirit*, which, in truth, is included in the introduction to the lectures on the *Philosophy of History* and thus belongs to the so-called mature stage of the system. Hegel recurs again to the example of the seed in order to show that the spirit is the subject that engenders itself by itself and accomplishes itself by exteriorizing itself in view of self-return:

> Spirit is essentially the result of its own activity: its activity is the transcending of immediate, simple, unreflected existence—the negation of that existence, and the returning into itself. We may compare it with the seed; for with this the plant begins, yet it is also the result of the plant's entire life. But the weak side of life is exhibited in the fact that the commencement and the result are disjoined from each other.[29]

Therefore, the seed and the plant describe a form of life that engenders and develops itself by itself, whose identity would be at the same time the beginning and the result, thus, an identity that does not need the other in order to produce itself and that is opposed to the other only from itself and in view of accomplishing itself as self-identity. If this form of natural life can only represent the life of the spirit without being it, that is precisely because of its natural limit: the plant produces a seed that is identical to the one that produced it but is another individual; there is no self-return here, whereas the spirit reproduces always itself and is the product of its production. This limit is properly the limit of nature; it affects also the natural life of the animal, and thus the human living as animal, and binds it to death. Later, it will be necessary to look at the transition that closes the *Philosophy of Nature* and announces the *Philosophy of Subjective Spirit*. For now I remark that Derrida singles out the distinctive traits that regulate the recourse to the example of the seed in the Hegelian system as follows—and, at the same time, he sheds light on the trace that one should follow in order to grasp the more profound reason of that recourse:

> The figure of the seed (let us call it thus provisionally) is immediately determined: (1) as the best representation of the spirit's relation to self, (2) as the circular path of a return to itself. And in the description of the spirit that returns to itself through its own proper product, after it lost itself there, there is more than a simple rhetorical convenience in giving to the spirit the name father. Likewise, the advent of the Christian Trinity is more than an empiric event in the spirit's history.[30]

In the *Philosophy of History*, as well as throughout the system, Hegel conceives of the Christian trinity as the highest representation of the life of the spirit realized by man before speculative dialectic: the infinite god—the father—posits himself in the finite—the son—and returns to himself without losing himself in the finitude—the death of the son. One should follow Derrida's long analysis; I limit myself to quoting just the conclusion that deduces the implications of the apparently metaphorical passage from the life of the plant to the one of the spirit: "The infinite father gives himself, by self-fellation, self-insemination, and self-conception, a finite son who, in order to posit himself there and incarnate himself as the son of God, becomes infinite, dies as the finite son, lets himself be buried, clasped in bandages he will soon undo for the infinite son to be reborn."[31] Therefore, a certain determination of the life of the plant permits us to think the infinite life of the determination of the spirit, through the representation of the trinitarian relation in Christianism. This was already at work, as Derrida remarks, at the beginning of the system, in particular, in *The Spirit of Christianity*, which, to follow the Hegelian logic at stake here, could be thus considered the germ of the system, even if the latter is not developed yet in all its parts. The botanic metaphor recurs three times and always to illustrate the relation of filiation between god the father and his son as infinite life. Let me quote the last occurrence:

> It is true only of objects, of things lifeless, that the whole is other than the parts; in the living thing, on the other hand, the part of the whole is one and the same as the whole . . . What is a contradiction in the realm of the dead is not one in the realm of life. A tree which has three branches makes up with

them one tree; but every "son" of the tree, every branch (and also its other "children," leaves and blossoms) is itself a tree. The fibers bringing sap to the branch from the stem are of the same nature as the roots. If a [cutting from certain types of] tree is set in the ground upside down it will put forth leaves out of the roots in the air, and the boughs will root themselves in the ground. And it is just as true to say that there is only one tree here as to say that there are three.[32]

It would be worth following the reading that highlights the role played by John's evangel in the context of this interpretation of the Christian trinity as infinite life. The values of life (*zoē*), light (*phos*), and truth (*aletheia*) are regularly associated in John's text. But at this point we may wonder where this determination of the life of the plant, as reproducing itself by itself, as identical to itself in its other, comes from. This determination that allows Hegel to define in those terms the immediate determination of natural and biological life in view of its speculative sublation in the infinite life of the spirit; that allows him to posit the life of the animals, and, thus, of man as a middle term, necessary to the operation of that sublation. The answer is: from biology. In the *Philosophy of Nature*, Hegel analyzes and translates into speculative terms the biology of his time. His practice in this regard remains unchanged from the early to the later writings, where we find also a description that recalls the examples of the overturned tree: "But there is no more familiar fact than that each branch and twig is a complete plant which has its root in the plant as in the soil; if it is broken off from the plant and put as a slip into the ground, it puts out roots and is a complete plant. This also happens when branches are accidentally severed from the plant."[33] Therefore, we should recognize in the Hegelian conception of the germ the legacy of the naturalistic theories of preformationism that were elaborated in the seventeenth and eighteenth centuries, and whose influence was still very strong in Hegel's time:

> The germ is the unexplicated being which is the entire Notion; the nature of the plant which, however, is not yet Idea because it is still without reality. In the grain of seed the plant appears as a simple, immediate unity of the self and the genus . . . The development of the germ is at first mere growth, mere increase; it is already in itself the whole plant, the whole tree, etc., in miniature. The parts are already fully formed, receive only an

enlargement, a formal repetition, a hardening, and so on. For what is to become, already is; or the becoming is this merely superficial movement. But it is no less also a qualitative articulation.[34]

According to preformationism, the adult animal, with organs and hereditary characters, is already present in miniature in the *germ*—that is, in the egg or the spermatozoon. The preformationism was contrasted by the theory of epigenesis according to which the embryo is developed on the basis of an *undifferentiated germ*, through the progressive formation of the various parts of the organism. In 1694, Nicolas Hartsoeker advanced the hypothesis that the whole fetus, a "homunculus" that is the microscopic duplication of the being in gestation, is located in the spermatozoon, with its encephalic extremity in the head of the spermatozoon itself.[35] Both preformationisms, *ovism* and *spermism*, are based on the relevance of reproduction for the study of living beings and share the theory according to which the adult animal is found already preformed in the germinal cells. Preformationism was first enunciated by the Dutch Jan Swammerdam, who, in a volume with a quite meaningful title, *Miraculum naturae sive uteri muliebris factorya* (1672),[36] denied that there is a true metamorphosis in insects. For Swammerdam, for instance, the butterfly is entirely present, with its organs being already distinguished, in the egg of the worm. He argued that all germs have existed since the beginning of the world insofar as Creation is a unique act. Therefore, in the moment of Creation, in the ovaries of Eve, there were already in miniature the human beings that are bound to be born up to the end of the world. The development of these beings merely consists in an *explication* (in Latin *evolutio*: *evolution*) of the parts packed in the germ, through successive, qualitative mutations (growing and enlarging). Among the followers of preformationism are included Leibniz, Bonnet, and Spallanzani, who are sources of *The Philosophy of Nature*. Moreover, the *Encyclopédie* of Diderot and d'Alembert deemed preformationism the most reliable hypothesis.[37]

A passage from "Force et signification" testifies to Derrida's knowledge of preformationism, which dates at least from 1963. Further, in this passage Derrida already recognizes the theologico-metaphysical presupposition of preformationism and, hence, the persistence of this presupposition in the "finalism" that represents the most refined theoretical development of that theory ("finalism" and, thus, of Kant and Hegel):

> By preformationism we indeed mean preformationism: the well known biological doctrine, opposed to epigenesis, according to

which the totality of hereditary characteristics is enveloped in the germ, and is already in action in reduced dimensions that nevertheless respect the forms and proportions of the future adult. A theory of *encasement* was at the center of preformationism which today makes us smile. But what are we smiling at? At the adult in miniature, doubtless, but also at the attributing of something more than finality to natural life-providence in action and art conscious of its works.[38]

Given that preformationism was invented to legitimate, within the life sciences, the unity of Creation in line with Christian dogma and that, ultimately, this theory rests on Aristotle's texts, it is possible to understand the complicated relations between Christian religion, philosophy, and biology that are at stake here and that are necessary to loosen in view of a deconstruction of the notion of life. In particular, in the context of the debate among naturalists, Hegel shares the position of Treviranus, for whom the reproduction of the plant does not imply sexual difference and, thus, consists in a pure self-reproduction, without difference or opposition to an other that is different from the self but of the same species as occurs in animal sexual reproduction: "This reproduction is not mediated by opposition, therefore it is not a unified emergence, although the plant can also rise to this. The emergence of true separation in the opposition of the sex relationship belongs to the power of the animal however."[39] Therefore, Hegel limits himself to speculate on what he receives from the life sciences of his time, even if, I note, he takes a precise position while being aware of the botanic theories for which the reproduction of the vegetal organism is a sexual reproduction. Goethe, whose *Metamorphosis of the Plants* is an essential source for Hegel, perhaps the most decisive for the section of the *Philosophy of Nature* dedicated to the vegetal organism, is a firm supporter of those theories. However, Hegel simply remains silent on this point. This passage is decisive for Derrida. In order to determine the self-reproduction of the spirit as the form of life that contains in itself the determinations of the natural life as its sublated moments, Hegel must affirm, against empirical evidence and scientific theories, that the reproduction of the vegetal organism does not go through sexual difference: "There would be no sexual difference in the plants."[40] However, when Hegel addresses the theories of sexual reproduction, in order to deny their legitimacy, the stakes of the position taken in the field of the botanic sciences appear evident:

(c) Now though this compels us to admit the occurrence of an actual fertilization, there still remains the third question, whether it is *necessary*. Since buds are complete individuals, and plants propagate themselves by stolons, and leaves and branches need only come into contact with the earth in order to be themselves fertile as distinct individuals (§ 345, Zus., p. 313), it follows that the production of a new individual through the union of the two sexes—generation—is a play, a luxury, a superfluity for propagation; for the preservation of the plant is itself only a multiplication of itself. Fertilization by sexual union is not necessary, since the plant organism, because it is the whole individuality is already fertilized on its own account even without being touched by another plant.[41]

Hegel seems to fall into the logic of the borrowed kettle, in which Freud recognized the symptoms of an incomplete repression. According to this logic, one supports contradictory arguments in order to affirm an unsustainable repression: 1) the reproduction of the vegetal organism does not imply a sexual difference, 2) the reproduction of the vegetal organism can imply a sexual difference, and 3) sexual differentiation is in any case superfluous for the reproduction of the vegetal organism. Whether we are facing a repression or a fidelity to a circular logic, we can find here the condition of natural life that must be repressed so that the infinite life of the spirit is accomplished as an imperishable self-reproduction without death: sexual difference, that is, difference, the relation to the other that irreducibly conditions the life of the living, and not only that of animals and of humans, the most evolved form that life obtains within the boundaries of nature. The final moment of *The Philosophy of Nature* unfolds the generic process in animal life and ends with the death of the individual. In order to reproduce itself, the animal individual needs to copulate with an individual of the other sex, and thus the product of reproduction is another individual different from the two generating it: there is no self-return in natural, sexual reproduction, no *Aufhebung*, no self-reproduction. Rather, there is dissemination, as Derrida himself suggests. Above all, it is worth remarking that the irreducible difference between the living individual and its concept, determined by sexual difference, also brings about the inborn death, as Hegel puts it, of the natural individual:

§375. Universality, in the face of which the animal as a singularity is a finite existence, shows itself in the animal as the abstract power in the passing out of that which, in its preceding process (§ 356), is itself abstract. The *original disease* of the animal, and the inborn *germ of death*, is its being inadequate to universality. The annulment of this inadequacy is in itself the full maturing of this germ, and it is by imagining the universality of its singularity, that the individual effects this annulment. By this however, and in so far as the universality is abstract and immediate, the individual only achieves an *abstract objectivity*.[42]

I quote the final paragraphs of Derrida's commentary:

> There is a natural death; it is inevitable for natural life, since it produces itself in finite individual totalities. These totalities are inadequate to the universal genus and they die from this. Death is this inadequation of the individual to generality; . . . Inadequation—classification and abstraction—of the generic syllogism: it has been demonstrated that inadequation placed in motion sexual difference and copulation. So sexual difference and copulation inhabit the same space; they have the same possibility and the same limit as natural death. And if the "inadequation to universality" is the "*original disease (ursprüngliche Krankheit)*" of the individual, as much ought to be able to be said of sexual difference. And if the inadequation to universality is for the individual its "inborn *germ of death (Keim des Todes)*," this must also be understood of sexual difference, and not only by "metaphor," by some figure whose sense would be completed by the word "death." *Germ of death* is almost tautological. At the bottom of the germ, such as it circulates in the gap [*écart*] of the sexual difference, that is, as the finite germ, death is prescribed, as germ in the germ. An infinite germ, spirit or God engendering or inseminating itself naturally, does not tolerate sexual difference. Spirit-germ disseminates itself only by feint. *In this feint*, it is immortal.[43]

We understand now why the dialectical sublation of the life of the animal organism is at the same time a recoiling, a repression of natural life, that is the irreducible condition of a finite living; why this sublation is a mise-en-scène, a feint. Above all, we understand which interest produced that scene: repressing différance as the irreducible condition of the life of the living, not just of its death but also of its life, as Hegel demonstrated when discussing sexual difference as the condition of animal reproduction, just before letting the curtain fall over natural life. Finally, to affirm that the life of the spirit is an infinite life, infinite self-reproduction, pure identity able to retain difference in itself, namely, death as a simple moment, Hegel hides or represses the possibility of the thought of a natural life, our life, which would account for différance as its irreducible condition of possibility. If we admit that différance is the irreducible, non-relievable condition of the life and death of the living, then we also understand the choice of the syntagm *"la vie la mort"* to allude to this dynamic of différance. This syntagm stands for recognizing différance at the heart of the life of the living. On the basis of *"la vie la mort"* we shall think another philosophy of life as well as another science of life, given that biology goes on to conceive of the logic of the living in terms of self-reproduction, as is the case in Jacob's book, which is ultimately given to reading and deconstruction.

III

The Absolute Programme

In the seminar *La vie la mort*, Derrida's reading of *The Logic of Life* is explicitly oriented to verify the hypothesis advanced in *Of Grammatology* in 1967: what are the deconstructive effects—if any—provoked by grafting the theory of information onto biological research and, in particular, by the use of notions such as "programme" and "writing"? Throughout the seminar, Derrida explicitly recalls this hypothesis (evoked again in *For What Tomorrow* in 2001) as well as the reasons that impose on it a critical vigilance and thus a verification:

> Some 10 years ago, in *Of Grammatology*, a chapter close to the beginning, entitled (just a coincidence, one would say, an almost subjectless prescience or teleology) *The Programme*, recalled that, I quote, "today the biologist speaks of writing and programme in relation to the most elementary processes of the information in the living cell." But this was not to reinvest in the notion or word of programme the entire conceptual machine of logos and of its semantics, but to try to show that the appeal to a non-phonetic writing in genetics had, would have, to imply and provoke an entire deconstruction of the logocentric machine rather than call for a return to Aristotle.[1]

Therefore, the recourse to the theory of information and in particular to notions such as "programme" and "writing" grants by itself neither the emancipation of biology from philosophy, the rigorous scientificity of biology, as Jacob believes, nor the deconstructive impact of biological discourse. Conversely, these notions can easily work at the service of the "logocentric machine" and thus of the metaphysical conceptuality that structures the Western philosophical tradition. To this extent, it would be possible to interpret the progress of genetic biology due to cybernetics within the framework of the tradition of the "philosophy of life," which finds its roots in Aristotle and has developed through Hegelian synthesis a tradition that genetic biology unconsciously inherits, being at the same time repetition and progress, a sort of evolutionary variation. This is precisely what Canguilhem argues in the essay "The Concept of Life," which is recalled by Derrida in this context:

> When we say that biological heredity is the communication of a certain kind of information, we hark back in a way to the Aristotelian philosophy with which we began . . . To say that heredity is the communication of information means somehow to acknowledge that there is a *logos* inscribed, preserved and transmitted in living things. Life has always done—without writing, long before writing even existed—what humans have sought to do with engraving, writing and printing, namely, to transmit messages. The science of life no longer resembles a portrait of life, as it could when it consisted in the description and classification of species; and it no longer resembles architecture and mechanics, as it could when it was simply anatomy and macroscopic physiology. But it does resemble grammar, semantics and the theory of syntax. If we are to understand life, its message must be decoded before it can be read.[2]

Derrida explicitly mentions this passage at the end of the first session of the seminar (1.22), after introducing his reading of Canguilhem's essay as an interpretation of contemporary biology in the light of the tradition of the "philosophy of life":

> "Philosophy of life," this is a quotation, in any case I use it here like a quotation. These are the last words of an article by Canguilhem entitled *The Concept of Life* and included in 1968 in *Etudes d'histoire et de philosophie des sciences*. I recommend you to read the article as well as the whole book, and *The Knowledge of life*, a previous work. . . . Philosophy of life, these are the last words of Canguilhem at the end of the article. They are not taken at his disfavor, and if the entire article is oriented towards the demon-

stration that contemporary biology is still profoundly Aristotelian and Hegelian this is not taken against him, the opposite is true.[3]

This passage allows us to point out in greater detail the contextual coordinates in which Derrida's seminar is inscribed, the reasons that led Derrida to treat Hegel as a key reference for both the metaphysical and the deconstructive significance of contemporary biology: the critical engagement with the thesis developed by Canguilhem in "The Concept of Life." From this perspective, the passage quoted by Derrida—corresponding to the last words of the article—is very important, as it raises a problem that will be essential for Derrida himself: if knowledge of life is immanent in the object studied—life—how should we understand their relationship?

> Knowledge, then, is an anxious quest for the greatest possible quantity and variety of information. If the *a priori* is in things, if the concept is in life, then to be a subject of knowledge is simply to be dissatisfied with the meaning one finds ready at hand. Subjectivity is therefore nothing other than dissatisfaction. Perhaps that is what life is. Interpreted in a certain way, contemporary biology is, somehow, a philosophy of life.[4]

To recall, I advanced the hypothesis that Canguilhem's article could be a source for these themes in *Of Grammatology*, and in particular in the aforementioned passage. This cannot be affirmed with certainty and would change nothing with respect to my previous work. Yet it is important to point out the following: 1) Derrida considers it necessary to verify the hypothesis advanced at the time of *Of Grammatology*, that is, why the recourse to the theory of information and in particular to the notion of "programme" and "writing" does not by itself necessarily entail deconstructive effects but, on the contrary, can also consolidate metaphysical sediments within scientific discourse; 2) this programme of verification must concern in particular the notions of "programme" and "writing." Therefore, what is at stake here is establishing on which bases the importation of these notions from cybernetics to biology and their specific use in biology can be justified. Respecting the programme of deconstruction, I start from the notion of "programme." Jacob ascribes to this notion a decisive role in the revolution that takes place in biology. According to Jacob, it is only thanks to the incorporation of the theory of information into the life sciences that it was possible to understand the role of DNA in the cell and, thus, finally, to describe the genesis and structure of genetic heredity on scientific bases:

> Even when the virtues of the scientific method had become solidly established for the study of the physical world, those who studied the living world continued to think of the origin of living beings in terms of beliefs, anecdotes and superstitions for several generations. Relatively simple experiments suffice to make short work of the notion of spontaneous generations and impossible hybridations. Nevertheless, some aspects of the ancient myths concerning the origin of man, of beasts and of the earth persisted, in one form or another, until the nineteenth century. Heredity is described today in terms of information, messages and code. The reproduction of an organism has become that of its constituent molecules. This is not because each chemical species has the ability to produce copies of itself, but because the structure of macromolecules is determined down to the last detail by sequences of four chemical radicals contained in the genetic heritage. What are transmitted from generation to generation are the "instructions" specifying the molecular structures: the architectural plans of the future organism.[5]

Hence only the importation of the theory of information would have allowed biology to become a true science—the model of which is represented by physics—and to set itself free from philosophy and the theological-metaphysical presuppositions sedimented in it. In particular, thanks to Aristotle, whose influence on the life sciences has been enormous and has extended far beyond the Renaissance, as Jacob acknowledges, through Kant's and, then, Hegel's recourse to teleology in the interpretation of the laws that govern nature in general and the living organism in particular:

> From ancient times to the Renaissance, knowledge of the living world scarcely changed. When Cardan, Fernel or Aldrovandus speak of organisms, they are more or less repeating what Aristotle, Hippocrates or Galen had already said. In the sixteenth century, each mundane object, each plant and each animal can always be described as a particular combination of matter and form. . . . The hand that confers form on matter to create stars, stones or living beings is that of Nature. However, Nature is merely an executive agent, an operative principle working under God's guidance. When one sees a church or a statue, one knows perfectly well that an architect or a sculptor exists or had

existed in order to bring those objects into being. In the same way, when one sees a river, a tree or a bird, one also knows that a supreme creative Power exists and, having decided to make a world, arranges it, keeps it in order and constantly directs it.[6]

The notion of "programme" allows us to understand the transmission of genetic heredity in a rigorously scientific fashion and, at the same time, to free teleology from its theological-metaphysical roots. Once it is recognized that the reproduction of the genetic programme constitutes the cause and end of the life of the living, it would no longer be necessary to refer to a divine intentionality as its condition. The end of life is reproduction and the laws of reproduction are inscribed in the living, in the genetic programme that regulates the construction of the organism in view of its reproduction and transmission in another living organism, generation after generation. It is no longer necessary to suppose a (divine) intentionality as the condition of the organized structure that establishes the living, or even an end that would be external and whose foundations would be beyond nature itself:

> In the chromosomes received from its parents, each egg therefore contains its entire future: the stages of its development, the shape and the properties of the living being which will emerge. The organism thus becomes the realization of a programme prescribed by its heredity. The intention of a psyche has been replaced by the translation of a message. The living being does indeed represent the execution of a plan, but not one conceived in any mind. It strives towards a goal, but not one chosen by any will. The aim is to prepare an identical programme for the following generation. The aim is to reproduce [*se reproduire*]. An organism is merely a transition, a stage between what was and what will be. Reproduction represents both the beginning and the end, the cause and the aim.[7]

Therefore, according to Jacob, once it is recognized that the reproduction of the genetic programme is the cause and end of the organism, it is possible to dissociate finalism from its theologico-metaphysical matrix and give it a scientific statute:

> In each case, reproduction acts as the main operative factor: on one hand, it provides an aim for each organism; on the other

it gives a direction to the aimless history of organisms. For a long time, the biologist treated teleology as he would a woman he could not do without, but did not care to be seen with in public. The concept of programme has made an honest woman of teleology [A cette liaison cachée, le concept de programme donne maintenant un statut légal].[8]

Therefore, if reproduction is "the main operative factor," the *telos* that orients the logic of the living, the genetic programme is the dispositive in charge of its realization.[9] As we shall see, Derrida subordinates the notion of reproduction to a subtle deconstruction, but, according to him, we must first reckon with the importation of the notion of programme from cybernetics to biology, with its specific use in Jacob:

What should we say of the value of programme that would finally be the solution of all problems, that would institutionalize, giving a *status* to it, . . . giving a legal statute, that is a scientific one, permitting it to be recognized as scientific, giving a legal statute to a discourse that otherwise would have been considered as non-scientific, metaphorical, ideological, imaginary or however you would like to call it, according to all the ways that one may have to determine the non-scientific, what has no right of citizenship within the scientific institution.[10]

In particular, for Jacob, the notion of programme imposed itself in the field of biology as it takes account of the two traits of the living that he considers immediately evident to common sense:

The concept of programme blends two notions which had always been intuitively associated with living beings: memory and design. By "memory" is implied the traits of the parents, which heredity brings out in the child. By "design" is implied the plan which controls the formation of an organism down to the last detail. Much controversy has surrounded these two themes.[11]

Jacob alludes to the controversy about the question of the "acquired characters" and in particular to the possibility that the external environment influences and modifies the genetic programme of a single individual by generating mutations that will be transmitted to successive generations.

For Jacob, this is a common illusion, a very old one, due to the ingenuous, intuitive and nonscientific use of the analogy between genetic memory and nervous memory (brain, mind, or psyche): "First, with respect to the inheritance of acquired characters. The idea that the environment can influence heredity represents a natural confusion between two kinds of memory, genetic and mental [*nerveuse*]."[12] The analogy is justified, but once it is used naively it can engender a dangerous confusion that leads us to identify genetic memory, the heredity that structures the living organism— to begin with its most elementary constituents—with nervous or cerebral memory, which specifies the behavior of the individual with respect to the environment. If we attribute to genetic memory the structural traits of nervous memory—acquisition, conservation, and transmission of the data that derive from the environment—then it will appear legitimate, supposing that we can intervene from outside to modify genetic memory, that is, the genetic programme. According to Jacob, the introduction of the notion of programme allows us to establish the correct use, that is, the effective, operative, and scientific use of the analogy between genetic memory and nervous memory, as it maintains the common characters and, at the same time, avoids the identification of the two systems of memory, so long as they correspond to two different and specific stages of the evolution of the living:

> For modern biology, the special character of living beings resides in their ability to retain and transmit past experience. The two turning-points [*points de rupture*] in evolution—first the emergence of life, later the emergence of thought and language—each corresponds to the appearance of a mechanism of memory, that of heredity and that of the mind [*cerveau*]. There are certain analogies between the two systems: both were selected for accumulating and transmitting past experience, and in both, the recorded information is maintained only as far as it is reproduced at each generation. However, the two systems differ with respect to their nature and to the logic of their performance. The flexibility of mental memory makes it particularly apt for the transmission of acquired characters. The rigidity of genetic memory prevents such transmission.[13]

Derrida draws attention to the analogy examined by Jacob and more generally to the very concept of analogy evoked here. First, for Derrida, Jacob imports a concept of philosophical provenance, such as "analogy," into biological discourse by attributing to it a decisive, operatory value but without the necessary critical accuracy with regard to the consequences that such a recourse would entail for the stability and consistency of a discourse that wishes to be rigorously scientific. The notion of analogy, of Platonic provenance, is in strict solidarity with the structure of hierarchized oppositions that characterize metaphysics (intelligible/sensible, signified/sign, proper/figured, concept/metaphor, and so forth); therefore, it works within this structure and, once adopted without any critical vigilance, cannot but confirm the whole system:

> As for the analogy in question, Jacob does not ask where it leads with its implications and with the very choice of its name. He determines that analogy as a similarity between two systems (in both cases, accumulation of a "past experience" and, in both cases, transmission of this experience). But one can only analyze the text to go further in the necessity and problematicity of this word analogy. Firstly, analogy is here between two systems and two logics, a system of relations of proportionality between multiple terms with variables. Just as nervous memory (that is, cerebral memory, thought and language in traditional sense) accumulates and transmits information, so does genetic memory. This relation, this relation of relations (among four terms) was called by the Greeks a *logos* and an analogy. Here the analogy between the two relations, between the two *logoi*, is a relation between a memory that involves language or logos in the current sense (nervous or cerebral memory corresponding to the second emergence) and a memory without language in current sense (genetic memory). Analogy in the logos of the modern geneticist (in his metalanguage or supposed metalanguage), between a logos in the so-called proper sense and an a-logos.[14]

For the deconstruction of the metaphysical presuppositions implicit in the notion of analogy, Derrida explicitly refers to the essay "White Mythology."[15] Here, in the seminar, he focuses on the consequences that the importation of the analogy carries within biological discourse. If what grants a properly scientific and non-metaphorical use of analogy is the discovery, realized by grafting cybernetics onto biology, that genetic memory works as a cerebral memory and thus as a language—that is, according to the structure and laws of the *logos*—and, therefore, if the *logos* is the guarantee of the rational structure (that is, of the logic) of the living, then Canguilhem is right to affirm that through cybernetics modern biology has unconsciously inherited the legacy of the philosophy of life from the Aristotelian matrix and, thus, of the metaphysical conceptuality on which the latter is grounded and from which Jacob wants to be free:

> But this general analogy has been only possible when (today) we got to know, with a scientific knowledge, that the a-logos was also a logos in the broad sense, that genetic memory operated *like* a language, with code, message, possible translation of message, and that it operated also by means of analogies, that is, putting relations into relations [*mises en rapport de rapports*], and more precisely by means of four radical elements.[16]

Once the relationship between genetic memory and cerebral memory has been subordinated to the order of *logos* and thus to the traditional conception of language that derives from it and still governs the function of the key features of the theory of information (programme, code, message, transmission), it risks importing into biological discourse, in its decisive articulation—the definition of the notion of programme—the logocentric structure that characterizes the metaphysical tradition from Plato to Hegel up to Saussure and beyond, as is demonstrated in *Of Grammatology*. In particular, it carries with itself the determination of the sign as the simple means of the external transmission of a signified constituted by itself, that is, produced within a certain ipseity (soul, subject, consciousness, brain) and, thus, autonomous and independent from the material exteriority of the signifier in charge of its transmission:

> Once this analogy is accepted without interrogating what is *logos*, a message and a code determined on the basis of their semiotic code, it is possible to ask if this is enough to make the subject disappear, what Jacob calls "the intention of a psyche," a formula that caricatures all traditional theological providences, in order to escape what the values of message, translation, design, end import from the system of *logos*, of traditional logocentrism.[17]

As I said earlier, the stakes of Derrida's reading consist in the possibility of liberating biological discourse from the bonds that tie it to the order of metaphysical discourse and to highlight the latter's deconstructive implications. As we shall see, these stakes are essentially related to the notion of writing developed by Jacob, to the role that this notion plays in the most precise and detailed determination of the logic that regulates genetic heredity and thus the life of the living. It is from this perspective that Derrida deepens the analysis of the analogy between genetic memory and nervous memory in order to show that the latter's precariousness and lack of rigor are the symptoms of the repression of an interpretative possibility that the adoption of the analogy itself prevents us from seeing for irreducibly structural reasons. According to Derrida, the analogy as formulated by Jacob is poorly constructed: in order to establish an analogy between two or more terms—entities or relations between entities—it is necessary that each term be constructed by itself and determined independently from the other. It is necessary that between the terms of the analogy we recognize a qualitative and essential distinction, a difference of nature, as Jacob argues with respect to the difference between genetic memory and nervous memory. However, the first determination of the analogy proposed by Jacob is quantitative and not qualitative: the two systems of memory differ from each other because of a greater or lesser flexibility or rigidity, and not because of their nature or essence. The consequence that Derrida draws from this is important and goes far beyond Jacob's intentions: "We are no longer dealing with two rigorously discontinuous types but with two relays of the same economy . . . therefore, the analogy is no longer an analogy between two different terms, but a similarity within the element of homogeneity."[18] If between genetic memory and cerebral memory there is a distinction that is only economical or quantitative, then not only is the analogy ungrounded, but it is also necessary to recognize that the two different articulations are within the same order: general memory, understood as the system of acquisition, conservation, and transmission that structures the genesis and evolution of the life of the living. In other words: différance—or arche-writing—as the structural condition of the history of life. Therefore, if they are phenomena of the same order, not only is it necessary to exclude their qualitative opposition, their difference in nature and essence, but it is also legitimate to hypothesize that between the two memories, a relationship is implied in their very determination, and thus the two systems are not each one the outside of the other, they are not determined independently from the other, but rather there is between them an evolutionary relationship, and thus the one—nervous memory—is a specific emergence in the course of the evolution of the living that is structured according to the conditions of the other—genetic memory. This means that to hypothesize memory as an individual psychic structure is an evolutionary product of genetic memory, of the programme that presides over the logic of the living and, ultimately, that between nature (genetic memory) and culture (cerebral memory) there is continuity rather than rupture, as the analogy established by Jacob supposes.

I do not want to promote here a rigidly deterministic biologism according to which any cultural phenomenon should be brought back to the biological-evolutionary conditions that structure the life of human beings and in particular of their brains. For Derrida, between genetic memory and nervous memory, it is possible to think of a differential relation and thus an articulation of the dynamic of différance as the general condition of the life of the living and of its evolutionary history:

> For my part, I would see no more than a progress in this suppression of a limit that has often served humanist or spiritualist ideologies or, generally speaking, the most obscurantist metaphysics. I would see no more than a progress here if the question of the logos of the analogy were elucidated in a critical fashion in order to avoid the return in force, merely legalizing a clandestine metaphysics, everything that has been attached to the value of logos and analogy across the tradition. To anticipate and to speak a little algebraically: I would be in favor of a de-limitation that destroys limits and oppositions (for instance, the two types of programme in which one would recognize on one side the pure genetic and on the other side the great emergence of the cerebral, from the being-erected to the *zoon logon ekon* and all that follows from it), destroying that opposition, then, not to give rise to something homogenous, but rather to a heterogeneity or a differentiality: for, as I was suggesting from the beginning, the functioning of the opposition has always had the effect of effacing differentiality. What interests me under the heading of the au-delà and of the pas au-delà, is precisely this limit without opposition of opposition and difference.[19]

Derrida seems to adopt an evolutionary perspective that is very close to that of Leroi-Gourhan, to whom, as we have seen, he is indebted for the elaboration of the notion of arche-writing in the framework of an evolutionary description of the genesis of the apparatuses of memorization—from the genetic programme to writing, through memory as the individual psychic structure. However, because for Derrida *différance* regulates the history of life, the evolution should be thought as a differential/differing process in which each stage of the process of differentiation corresponds to a specific, relatively independent emergence that consists in an articulation and an effect of differential iteration with respect to the conditions of the process itself. Therefore, for Derrida, evolution is not a linear and continuous (whether teleological or not) process and does not require leaps and irreducible ruptures. The hypothesis adopted here is neither "deterministic" nor "reductionist," not even metaphysico-humanist and logocentric: between the genetic and the symbolic, between nature and culture there is neither identity nor opposition but différance.

On the other hand, the hypothesis advanced here should not surprise a careful reader of Derrida. It has to do once again with the *programme* elaborated in the essay "Différance":

> Thus one could reconsider all the pairs of opposites on which philosophy is constructed and on which our discourse lives, not in order to see opposition erase itself but to see what indicates that each of the terms must appear as the différance of the other, as the other different and deferred in the economy of the same (the intelligible as differing-deferring the sensible, as, the sensible different and deferred; the concept as different and deferred, differing-deferring intuition; culture as nature different and deferred, differing-deferring; all the others of *physis—tekhne, nomos, thesis*, society, freedom, history, mind, etc.—as *physis* different and deferred, or as *physis* differing and deferring. *Physis* in différance.[20]

It is to this programme that Derrida refers again much later, in *For What Tomorrow* (2001), in an important passage in which he explains his position on the alternative between "naturalism" and "constructivism":

> I would prefer not to let myself get trapped in an alternative between naturalism and constructivism. And I do not consider legitimate any of the numerous conceptual oppositions evoked, presupposed, or taken as firmly established in such an alternative. I try to be neither a naturalist nor a constructivist—particularly if the latter refers to some sort of totally deracinated artifactual confection, outside of all biological premises. . . . In all these

problems, which are considerable, I would not want to renounce either side. I would like to find a way to take into account biological and genetic determinisms, which are themselves complex and not simply "natural." In the phenomena of biology and genetics there are encodings, directed bifurcations, "languages," and modes of "writing." Put another way, there is a sort of "culture," even a "technique" of genetics that makes possible all sorts of constructions. So I don't want to renounce biological and genetic knowledge insofar as it has something open about it, something progressive and perfectible. However, the *psyche*—or culture, or the symbolic, to take up, without accrediting them, these equivalences so often taken for granted—the *psyche*, then, takes over where the so-called genetico-biological laws leave off, precisely in a way involving difference. At certain "moments," this difference can interrupt these laws; at other moments, it can introduce the economy of a new configuration into the immanence of the living being. The interruption itself belongs to the field of what is genetically or biologically possible. These are not only different "moments" of difference. Difference means at once *the same* (the living being, but deferred, relayed, replaced by a substitutive supplement, by a prosthesis, by a supplementation in which "technology" emerges) and *the other* (absolutely heterogeneous, radically different, irreducible and untranslatable, the aneconomic, the wholly-other or death). An interruption involving difference is both reinscribed into the economy of the same and opened to an excess of the wholly other. To return to this word, there is some *psyche*, that is, there is some "life," as soon as this difference appears, or more precisely (for it may not appear *as such*, no doubt it never does) as soon as it leaves a *trace* (neither a sign nor a signifier nor anything whatever that one might call "present" or "absent," but a *trace*).[21]

The way in which Derrida speaks of Freud in this passage suggests an explicit reference to the seminar *La vie la mort*: "Between the two [naturalism and constructivism], you inscribe the concept of 'psychic life.' Here, too, we would need to know what is meant by *psyche*. In Freud, the relation of the psychical to the biological is, as you know, always suspended, set aside to be worked out later, in future generations, and it is therefore in truth very complicated."[22] To support his position, Derrida recalls Freud's attempt to elaborate a "metapsychology," that is, a speculative theory that would account for the articulation of the biological and psychic according to the coordinates established in "Freud and the Scene of Writing" and developed, as we shall see, in the part of the seminar dedicated to *Beyond the Pleasure Principle*: "Freud never abandoned the idea of finding a biological foundation for psychic organization, which did not prevent him from renouncing the construction of a 'biology of the mind.' On the contrary, he turned toward the construction of a metapsychology, distinct from classic psychology, which aimed at elaborating theoretical models not directly tied to clinical observation."[23]

I trace this "differential" thesis throughout the seminar and beyond in order to verify its sustainability and implications within a deconstructive perspective. For the moment, I go on with the analysis of the notion of programme and the deconstruction of the analogy between genetic memory and cerebral memory. Derrida takes into consideration another distinctive trait, this time of a qualitative nature, which seems to be more rigorous and thus able to legitimate the opposition between the two systems of memory and the very possibility of establishing a relation of analogy. For Jacob, genetic memory and cerebral memory differ in their relation to the outside: while the cerebral memory is open to the outside and is subject to its modifications, genetic memory would be impermeable to external action: "The genetic programme, indeed, is made up of a combination of essentially invariant elements. By its very structure, the message of heredity does not allow the slightest concerted intervention from without."[24] If we look more closely, this is not the case: affirming that genetic memory is impermeable to the action of environment would mean denying the possibility of selection, which is essential to evolution. For Jacob, the two systems are not opposed because of their opening or closure to the environment; rather, they are both open to the outside. Their opposition concerns the relation that they entertain with the environment: the cerebral memory interacts in a direct, conscious, and intentional way, with deliberate aims, and thus can modify its behavior, while the modifications genetic memory undergoes—the so-called mutations on which natural selection and thus the environment intervene ex post facto—would be of the order of contingency, accidental and deprived of a direct cause/effect relationship:

> Whether chemical or mechanical, all the phenomena which contribute to variation in organisms and populations occur without any reference to their effects; they are unconnected

with the organism's need to adapt. In a mutation, there are "causes" which modify a chemical radical, break a chromosome, invert a segment of nucleic acid. But in no case can there be correlation between the cause and the effect of the mutation. Nor is this contingency limited to mutations alone. It applies to each stage in the formation of an individual's genetic heredity, the segregation of the chromosomes, their recombination, the choice of the gametes which play a role in fertilization and even, to a large extent, to the choice of sexual partners. There is not the slightest connection between a particular fact and its consequences in any of these phenomena. Each individual programme is the result of a cascade of contingent events. The very nature of the genetic code prevents any deliberate change in programme whether through its own action or as an effect of its environment. It prohibits any influence on the message by the products of its expression. The programme does not learn from experience.[25]

However, Jacob himself is forced to admit in another point of the text that, even if indirectly, the genetic programme does "learn from experience." The mutations of the programme, because of contingent events, must always undergo natural selection, which favors those mutations that adapt better to the environmental conditions that influence their possibility of reproduction:

In the end, the text is always rectified. But it is rectified neither by a mysterious will seeking to impose its design, nor by an environmentally determined reordering of the sequence: the nucleic-acid message does not learn from experience. The message is rectified automatically by a process of selection exerted, not on the genetic text itself, but on whole organisms, or rather populations of organisms, to eliminate any irregularity. The very concept of selection is inherent in the nature of living organisms, in the fact that they exist only to the extent they reproduce. Each new individual which by mutation, recombination and addition becomes the carrier of a new programme is immediately put to the test of reproduction. If this organism is unable to reproduce, it disappears. If it is able to reproduce better and faster than its congeners, this advantage, however minor, immediately favours its multiplication and hence the propagation of this particular programme. If in the long run the nucleic-acid text seems to be moulded by environment, if the lessons of past experience are eventually written into it, this occurs in a roundabout way through success in reproduction.[26]

For Derrida, this opposition is again not rigorous enough: as psychoanalysis and modern structural sciences—such as linguistics, semiotics, anthropology, and so forth—have demonstrated, it is also possible to extend to cerebral memory and thus to the sphere of language, the sphere of the symbolic and of culture more generally, what Jacob maintains as an exclusive, distinctive trait of genetic memory. Also, cerebral memory, the individual psyche, is strongly bound to codes and programmes (linguistico-semiotic, social, religious, politico-institutional, economic, and so forth) in relation to which the margin of the intentional and deliberate intervention is very tight and aleatory: the programme remains far beyond the threshold of individual consciousness and thus of its possibility of action and deliberation. Consciousness is indeed an effect rather than a cause. As in the case of Jacob himself, who, in order to define the genetic programme, must resort, against his own conscious intention, to the most traditional philosophical tools, from which he believes he has emancipated himself:

> The heterogeneity of causes and effects, the non-deliberate character of changes in programme, in a word, all that places the subjects within the system in a situation of unconscious effects of causality, all that produces effects of contingency between the action coming from the outside and the internal transformations of the system, all of that characterizes the non-genetic programme as much as the genetic one. Where does Jacob get the idea that outside of the genetic system and programme the change of programme is deliberate, essentially deliberate? Where, if not in a metaphysico-ideological opposition that determines the superior and symbolical programmes (with humanity at the highest level) on the basis of sense, consciousness, freedom, knowledge, the limit between the inside and the outside, objectivity and non-objectivity, etc. Now, if something has been achieved by the so-called structural sciences today, it is the possibility of affirming that the systems related to language, the symbolic, cerebral memory, etc. also have an internal functioning, with an

internal regulation that escapes deliberation and consciousness and makes the effects come from the outside be perceived as contingencies, heterogeneous forces, which is necessary to interpret, translate, assimilate into the internal code, attempting to master them in that code, or failing to do so to the point that "mutations" are produced that can take on all sort of forms but which always signal a violent intrusion from the outside forcing a general restructuration.[27]

Therefore, Derrida seems to defend an ultra-deterministic thesis by attributing to symbolical programmes the very rigidity of the genetic programme claimed by Jacob. However, the opposite is true, as the apparently paradoxical conclusions of the argument prove: the two programmes—the genetic programme and the symbolic one—function on the basis of different principles of internal regulation, in view of their reproduction; yet this rigidity does not exclude their opening onto the outside; rather, it implies for both systems the possibility of being influenced and modified by what comes from outside and thus the necessity of interpreting what comes from outside with respect to the exigencies of the reproduction of the system. What comes from outside can simply be rejected from the programme if it is interpreted as dangerous; it can be assimilated, conserved, and thus also transmitted if it is interpreted as useful to the survival of the system; it can induce corrections in the mechanisms in charge of the execution of the programme; ultimately, it can induce modifications of programme and thus true mutations. This works for the symbolic programmes as much as for the genetic programme, with effects that both systems cannot easily control, as they are relatively aleatory, to the extent that they are opened onto the outside and called to interpret its contingency. The thesis implicit in Derrida (for whom there is a genetico-differential relation between genetic memory and nervous memory that results from différance as the common condition of their emergence and specific articulation, through different levels of development) contradicts one of the fundamental principles of the biology of the time, formulated by Jacob in the aforementioned passage: "The programme does not learn from experience," that is, genetic mutations cannot be caused by the environment where an individual lives.

At the time of its elaboration, Derrida's position would have been liquidated as an ingenuous Lamarckism, and this is perhaps one of the reasons why the seminar was not published. But the state of the art in biology is much different today; in particular, research that is considered of revolutionary relevance has been carried out and is congruent with Derrida's thesis insofar as the latter could be read as an anticipation and legitimation of this research from a theoretical point of view. I allude to the most recent research conducted in the field of *epigenetics*, a science that studies the interaction between genes and the environment, whether internal (the cellular environment) or external (what we ordinarily understand as the environment). This research has provoked a radical mutation of the order of biological discourse and thus of the interpretation of the logic of the living and of its evolution. In particular, the role of the genetic programme in the construction of the architecture of the individual appears today less deterministic and, ultimately, less exclusive. The architecture of the individual is no longer considered the exclusive result of the rigid execution of the genetic programme in its cells; rather it hinges on the interactions between genes and the cellular environment in which the genes are inscribed and on which they depend for the expression of their function. In particular, this expression undergoes a series of epigenetic regulations (methylation, RNA interference, histone modifications, genomic imprinting) that in some cases may depend on environmental factors external to the individual, such as pollution or a shortage or excess of food, but also to factors of psychological stress within the social or cultural order, such as insufficient elder care or traumas resulting from war. Some epigenetic regulations can even provoke a reassembling of the genetic programme of the individual ("reprogramming"), a reassembling that in some cases can be transmitted to the following generations and thus become hereditary. This feature is obviously decisive, not only because it allows us to affirm, against Jacob and with Derrida, that "[t]he programme learns from experience" but also because, from a more general perspective, it legitimates the hypothesis that these epigenetic regulations are essential factors of evolution, that is, of those genetic mutations that until now have been generically attributed to selection, which, as we saw above, intervenes on entire populations and not on individuals. In other words, it legitimates the hypothesis that those genetic mutations registered on the scale of populations are not only and exclusively due to mistakes in the transcription of the genetic programme, which would be independent from the environmental factors that intervene only in the selection of more adaptive mutations. Indeed, these mutations may be epigenetic adaptations in singular individuals exposed to specific external or internal environmental factors. To strengthen my argument and to grasp a deeper understanding of the relevance of this research, it may be useful to quote from the beautiful synthesis provided by Richard

C. Francis in *Epigenetics: The Ultimate Mystery of Inheritance*.[28] In particular, Francis focuses on the change of perspective induced by epigenetics with respect to the rigidly deterministic conception of the genetic programme, which had been endorsed for a long time after the studies carried out by Monod and Jacob. Francis considers this conception as the modern reformulation of the old preformationism in which, as we saw, Derrida found a metaphysical legacy:

> On the new, genetically inspired preformationist account, your genes contain the complex form that is you, through which they direct the process of your development. Genetic preformationism was successfully packaged through a series of intuitively appealing metaphors. First came the metaphor of the "genetic blueprint," then that of the "genetic recipe," and finally the "genetic programme." Some combination of the recipe and programme metaphors remains popular; they have in common the notion that the genes provide instructions, which cells execute. In genetic preformationsim, the executive gene is scaled up to become the executive genome . . . Whatever their intuitive appeal, these metaphors cannot withstand even the most cursory scrutiny. You couldn't cook up a single cell, much less a human being, given the instructions in the genetic recipe. Much of what you need to know lies elsewhere. More to the epigenesist point, most of the information in the recipe that goes into making you is not there from the outset. Rather, development is the process whereby this information comes to exist. The recipe is written during development, not prior to development. The same goes for the metaphor of a genetic programme. But the notion of a genetic programme also suffers from another fatal flaw: the software/hardware distinction. The genes are supposed to comprise the software and the rest of the cellular constituents the hardware whose operations the genes instruct. But as we have seen throughout this book, our genes are as much part of our hardware as any other biochemicals, and as much instructed as instructors. In fact, the sciences of epigenetics makes sense only when genes are viewed as biochemical hardware.[29]

Similar accusations of preformationism apply to more recent interpretations according to which the epigenetic regulations would be essential for the execution of the genetic programme but are still tied to the programme that would conserve the role of exclusive direction in its execution:

> The genetic-epigenetic programme metaphor acknowledges the central role of epigenetic events in the development but views them through a preformationist lens. In essence, the idea is that the epigenetic events described earlier are programmed by the executive genome. All of the problems with the "programme" metaphor (as in "genetic programmes") apply to the notion of a genetic-epigenetic programme—plus an additional one: in what sense are these epigenetic events programmed? Certainly not in the sense of "programme" that most readers will bring to the table: a recipe-like set of instructions. As we have seen, the epigenetic changes in gene expression that determine a cell's fate are largely determined by the position of that cell in the developing embryo. Therefore, it would be apt to say that the genes are programmed by cellular interactions.[30]

Therefore, however important it may be, the genetic programme is only an element of the cellular interactions that determine cellular differentiation and the organization of the living. Not only does it no longer represent the exclusive principle that regulates the rigid architecture of the living, but it is also subject to mutations due to the internal and external environments:

> Some epigenetic alterations of gene behavior have effects that extend beyond an individual lifetime. The effect of these transgenerational epigenetic alterations may be direct or indirect. Direct transgenerational effects occur when the epigenetic mark is transmitted directly from parent to offspring, through sperm or egg. This is what I call "true epigenetic inheritance." True epigenetic inheritance is not common in mammals like us, but it does occur. Indirect transgenerational effects are much more common . . . Much more indirect are the transgenerational effects observed in the maternal behavior and stress response. Here, the epigenetic alterations that influence these behaviors are re-created by through the social interactions that they both influence and are influenced by. This transgenerational effect is a positive feedback loop involving gene action and social

interaction. Whether direct or indirect, these transgenerational epigenetic effects should expand our notion of inheritance.[31]

It may be interesting to look more closely at the elements of congruency between the theses proposed by Derrida about the opening of the genetic programme and the results of the most recent epigenetic discoveries. Here it is worth remarking that the deconstruction of the notion of "programme" elaborated by Derrida does not contradict the contemporary perspectives of biological research, and thus it is possible, on the basis of this congruency, to prepare the way for closer debate between deconstruction and biology. Furthermore, what has been said seems to me sufficient to undermine Catherine Malabou's thesis that Derrida's elaboration of "arche-writing" would be closely bound to the concept of "programme," elaborated by cybernetics and implemented by biology, and thus that it could no longer be retained today, as it has become obsolete.[32] However, the traces of the deconstruction of the programme and, in particular, of its deterministic features can already be found in published texts such as *Dissemination*: "As the heterogeneity and absolute exteriority of the seed, seminal différance does constitute itself into a programme, but it is a programme that cannot be formalized. For reasons that can be formalized. The infinity of its code, its rift, then, does not take a form saturated with self-presence in the encyclopedic circle. It is attached, so to speak, to the incessant falling of a *supplement to the code* [d'un supplément de code]."[33]

At this point, the aim of Derrida's analysis is clear: the definition of the notion of programme, as it is formulated by Jacob, imported by cybernetics and transplanted at the heart of the living, is unconsciously overdetermined by the programme of metaphysics, with its fundamental logocentric and humanistic legacy: "Here too, you can see, the opposition between the two programmes cannot be rigorous, and this seems to me to depend upon the fact that, in absence of a reelaboration of the general notion of programme and the value of analogy, they remain marked by a logocentric teleology and by a humanist semantics, by what I would call a philosophy of life."[34] And it is exactly at this point of the seminar that we find the reference to Canguilhem that we took as a point of departure. Canguilhem's hypothesis—that modern genetic biology is the heir of the tradition of the philosophy of life—seems to be shared by Derrida but from a negative perspective. This does not mean that Jacob's enterprise and, more generally, the significance of genetic research is disqualified in a definitive way. In view of relaunching the stakes from a deconstructive perspective, it is necessary to reconsider Jacob's discourse with a critical and careful vigilance, to see to what extent it is possible to liberate biology from the uncanny specter of metaphysics that haunts it from within. As we shall verify, the outcome of this operation is bound to another notion that should be examined according to the programme established by Derrida: the notion of writing to which Jacob recurs in order to describe in detail the functioning of hereditary transmission. But before following this path, we should insist on the unconscious metaphysical legacy that has been uncritically incorporated into biological discourse, and, in particular, we should insist on its epistemological consequences. For Derrida, as we shall see, on these points rests the legitimacy of the philosophical investigation of the living and thus the necessity of thinking another relation between science and philosophy, against the latter's supposed liquidation from the life sciences.

The Essence of Life: Between Philosophy and Science

From the previous analyses, we may argue that Derrida calls into question the emancipation of the biology of his time from philosophy, as well as the scientific rigor that, through the notion of programme, the biological discourse would have gained against the philosophical one, which was considered nonscientific. Ultimately, he calls into question the rigor of these discriminating distinctions between biology and philosophy, of the opposition between scientific discourse and nonscientific discourse. This holds at least for Jacob, who never thematizes in a rigorous and scientific fashion the critique of the philosophical presuppositions of the life sciences, through which the alleged emancipation would have been accomplished; rather, he limits himself to generic references to the history of philosophy and the obscurantism of the religious matrix. Derrida often refers to the limits of Jacob's discourse in relation to these epistemological problems. He develops a deconstruction of these discriminating oppositions along the double register that informs the first sessions of the seminar, which I isolate here from their context in order to highlight their sense and implications. On the one hand, Derrida gathers together the symptoms of the metaphysical sedimentations at work in Jacob's biological discourse that were neither recognized nor thematized as such. On the other hand, he focuses on the metaphorical tenor of certain notions and arguments to which Jacob recurs in the framework of a discourse that wants to be rigorously scientific. He begins with the notion of "programme," where, as we have seen, the two registers of Derrida's reading are perfectly articulated with one another. Through the analysis of the notion of programme, Derrida had already drawn attention to the first symptom of Jacob's naiveté apropos the supposed emancipation of biology: "Despite his emancipation from philosophy, at least the emancipation we grant to these modern biologists, today, rather than the one they granted to themselves, Jacob regularly says 'beings' to designate the livings and 'things' to designate the non-livings."[35] Derrida does not overestimate the significance of this double opposition—being/thing and living/non-living—that is evidently imported from ordinary language, perhaps, in order not to credit another philosophical presupposition inherited from the tradition and still operative in Heidegger: the sovereign privilege of philosophy as general ontology before the other sciences that would hierarchically depend on it as regional ontologies.[36] At the same time, Derrida insists on the problems that the naive use of these notions entails for a biological discourse that wants to be rigorously scientific:

> Under this loose way of writing and this concession to current language that makes the living into "beings" and the non-living into "things" is hidden an enormous sedimentation I do not want to uncover here, any more than I want to give the impression that I am hounding a scientist in the exercise of his discourse on the basis of a demand of rigor coming from an ignorant philosopher. But I believe that this kind of indices must be taken seriously and reveal in the scientist who writes more philosophical *and* scientific naiveté than one generally dares to assert.[37]

There is first a problem of epistemological order related to the constitution of the fields of scientific research: those sciences that aim to have a rigorous theoretical foundation must not overlook the philosophical sedimentations unconsciously inherited through ordinary language and tradition. In particular, this holds for Jacob, who had found in a series of prejudgments the cause of biology's belatedness in relation to physics and, above all, of the influence to which biology was subjected in the rigorous determination of its own object:

> As everyone knows, it is in biological science that the nonscientific (conveyed [*véhiculé*] by current language or by philosophical language, which are often the same) contaminates

the very position of scientific problems from within. And this happens for essential reasons. Indeed, Jacob basically recognizes it when he admits that the delay in scientificity is more regular in the studies of the living than in those of the non-living.[38]

Derrida identifies the being/living association within biological discourse as a symptom of the possible sedimentation of the metaphysical matrix that, as he suggests from the beginning of the seminar, determines the different philosophical conceptions of life developed throughout the Western tradition, in particular, the variants influenced by Christianity, such as the more than emblematic case of Hegel's philosophy: "Calling the livings [*les vivants*] beings, as we do it in an ordinary language marked by a culture that is philosophical and Christian at once, which makes what lives and speaks into being [*qui fait être ce qui vit et parle*], here you have what is in accordance with our earlier remarks about the equivalence of being and living with all its problematic centre."[39] Once uncritically absorbed, this metaphysical sedimentation risks contaminating biological discourse with effects that could be destructuring, not only because it threatens the alleged emancipation of biology from philosophy but also because it necessarily limits the groundbreaking effects of the biological research of the time by tying them to a metaphysical matrix so powerful and pervasive that it passed through the ordinary language of tradition without being noticed. Evidently, Derrida is not interested in restoring the primacy of philosophy over other sciences, or over biology in particular. Rather, in the very interest of scientific research, he wants to draw attention to the necessity of a rigorous critical vigilance toward the philosophical sedimentations that we can find in scientific discourse, and thus to the necessity of deconstructing the metaphysical matrix that determines Western discourse about life:

> I would like to propose, not to reread with you *The Logic of Life*, but to construct with this book (or this discourse or the discursive ensemble of whom it is today an eminent and let's say highly concentrated representative), to reconstruct the machine by which it is governed, clearly without knowing it, "without knowing" meaning not only that its author does not know—this is too evident on every page of the text—but also that the system does not know. By this I do not mean that the system should be aware or not of being governed by that machine (I do not know what that would mean) but that the system does not

exhibit, and does not put into motion certain relations among statements or textual functions, and thus is not constructed with the maximum of power and efficacy, with the power and efficacy that I believe to be possible today.[40]

According to Jacob, by uncovering the physico-chemical unity of the processes of the organization of matter at a molecular level, thermodynamics allows biology to demarcate itself from philosophy, to abandon the concept of "life" understood as "metaphysical entity,"[41] as the abstract essence back to which the organization of living systems should be retraced. Biology today no longer needs to speculate philosophically on the essence of life; rather it analyzes living systems in a scientific way, that is, with the means provided by physics and chemistry. Therefore, biology replaces the abstraction of speculation with experimental research on the living:

> Recognition of the unity of physical and chemical processes at the molecular level has deprived vitalism of its *raison d'être*. In fact, since the appearance of thermodynamics, the operational value of the concept of life has continually dwindled and its power of abstraction declined. Biologists no longer study life today. They no longer attempt to define it. Instead, they investigate the structure of living systems, their functions, their history. Yet at the same time, recognition of the purpose [*finalité*] of living systems means that biology can no longer be studied without constant reference to the "plan" of organisms, to the "sense" which their very existence gives to structures and functions.[42]

For Derrida, this change of perspective is not sufficient to prove that biology is emancipated from philosophy. It is necessary that the experimental analysis of concrete living systems presuppose a definition or at least a pre-comprehension of what allows us to recognize a system empirically as a living system. A definition of the life of the living, even if implicit or unconsciously received from common sense, is required in the experimental analysis to which the living is submitted. Hence, abandoning the concept of life as a metaphysical hypostasis does not mean abandoning the ground of the philosophical inquiry into the essence of the living:

> Jacob cannot do without a reference to the essence of the living and he does it massively. It is not enough to replace "life" with

> "living" to escape the philosophico-socratic question: what makes the living a living? You speak of a living: then you should know or seek to know what you mean by that, by the being-living of a living, by the livingness [*vivance*] of the living, in other words, by the life of the living, the difference between the living and the non-living. And if you are right not to want to make life into an entity or an abstract and separated essence, you cannot do without implying that living means something and that there is a being-living of the living, a livingness or a life, which is precisely the very thing you are studying. What modern science has *perhaps* transformed is the concept of this essence of life, of the being-living, but not at all the reference to an essence of the living as such.[43]

For Derrida, it is possible to find in Jacob's discourse this reference to essence in the definition of "re-production" as an "operative factor" that regulates the logic of the living: "In fact, Jacob posits and defines it very frequently, it is a leitmotiv of his book: a living is recognized by its capacity to reproduce itself."[44] The essence of the living is reproduction, or, more precisely, as Derrida points out, self- or auto-reproduction: "For Jacob always says re-production where he is clearly describing a self-reproduction: there are non-livings that re-produce without re-producing themselves and this flexion onto themselves, this auto-affection is an essential fold of the structure."[45] As we shall see, this addition by Derrida is very significant because it points to a complication that the notion of "reproduction" cannot elude once it is understood as self- or auto-reproduction. For the moment, I only observe that, for Derrida, Jacob's discourse is inconsistent and contradictory, as he maintains that, on the one hand, on the basis of the unity of the physico-chemical processes at work in matter, there would be no change in essence between matter and the human being, whereas, on the other hand, self-reproduction would consist in the essential property of the living as such:

> And yet biology has demonstrated that there is no metaphysical entity hidden behind the word "life." The power of assembling, of producing increasingly complex structures, even of reproducing, belongs to the elements that constitute matter. From particles to man, there is a whole series of integration, of levels, of discontinuities. But there is no breach either in the composition of the objects or in the reactions that take place in them; no change in

"essence." So much so, that investigation of molecules and cellular organelles has now become the concern of physicists. . . . This does not at all mean that biology has become an annex of physics, that it represents, as it were, a junior branch concerned with complex systems. At each level of organization, novelties appear in both properties and logic. To reproduce is not within the power of any single molecule by itself. This faculty appears only with the simplest integron deserving to be called a living organism, that is, the cell. . . . The various levels of biological organization are united by the logic proper to reproduction.[46]

Strictly speaking, if self-reproduction consists in the property of the living and only of the living, how can we claim that there would be no change in essence between molecular particles and the human being? Although it depends on the physico-chemical processes detected in matter at the molecular level, self-reproduction is an absolute novelty with respect to matter, which is unable to reproduce itself, and it is this novelty that physics and chemistry could not explain before the arrival of information science—so it was, at least, for Jacob. However, also in this case, the question is not so much reestablishing the primacy of philosophy before the sciences and, in particular, biology, as uncovering in the definition of self-reproduction as the essence of the living the possible metaphysical sedimentation that biological discourse unconsciously absorbed. In fact, the definition of self-reproduction as the essential determination of the living does not by itself constitute a novelty introduced by biology once it is promoted as a rigorous science. We find it already affirmed in Hegelian philosophy, which, as we have seen, Derrida identifies with the most accomplished modern version of the metaphysical determination of life: "As if by chance, the essential definition is given of livingness [*vivance*], of what makes it that an existence (a living system or individual) is living, is literally the definition that the most metaphysical of the metaphysicians, the metaphysician *par excellence*, Hegel, gives of it, that the living individual is living insofar as it can reproduce itself."[47] Once again, Derrida refers to the treatment of "life" as the first moment of the absolute Idea, the apex of the *Science of Logic*. Reproduction amounts to the third moment of the dialectical determination of the "living individual," the *Aufhebung* of "sensibility" and "irritability":

> The two first moments, sensibility and irritability, are abstract determinations; in reproduction life is *something concrete* and vital;

in it alone does it also have feeling and power of resistance. Reproduction is the negativity as simple moment of sensibility, and irritability is only a vital power of resistance, so that the relation to the external is reproduction and identity of the individual with itself. Each singular moment is essentially the totality of all; their difference constitutes the ideal determination of form which is posited in reproduction as the concrete totality of the whole. On the one hand, therefore, this whole is opposed to the previous determinate totalities as a third, namely as a *concretely real* totality; on the other hand, however, it is their implicit essentiality and also that in which they are comprehended as moments and where they have their subject and subsistence.[48]

According to its universal determination, the living is an immediate identity with itself, "sensibility," namely, self-feeling—what Derrida calls "auto-affection"—but to be itself the living must posit itself as such, that is, it must make this determination objective, concrete; therefore, it must relate itself to an alterity in which it must be realized according to its determination; it must differentiate itself from itself, being other. "Irritability" stands for the disposition of the living to suffer alteration from the alterity to which it relates in order to be itself and offers resistance. It is only through reproduction, that is, through the production of another individual (that is different as individual but identical in structure), that the living concretely realizes the dialectico-speculative identity of the immediate universal identity and the particular difference, of being itself and being other than itself. Derrida does not limit himself to highlighting a mere resonance between Hegel and Jacob apropos reproduction as the essence of the living. To prove that *The Logic of Life* rests on the definition of the essence of the living and that this definition in turn refers to a powerful metaphysical sedimentation, he draws together a series of quotations from Jacob's text in which self-reproduction, even if it is never literally defined as the essence of the living, fulfills that function. Keeping in mind the passages already quoted, here I recall the following selection: "The living being does indeed represent the execution of a plan, but not one conceived in any mind. It strives towards a goal, but not one chosen by any will. The aim is to prepare an identical programme for the following generation. The aim is to reproduce [*se reproduire*]."[49] "Everything in a living being is centered on reproduction [*tout est agencé en vue de la reproduction*]."⁵⁰ "In such a system, however, reproduction, which is the very cause of existence, also becomes its purpose [*fin*]."⁵¹ "'reproduction,' is the intrinsic property of all living systems."⁵² Therefore, self-reproduction is (1) the internal property of the living, (2) what the living aims for, and (3) the cause and end of the living. On the basis of these occurrences, not only is it justified to affirm that, no matter how consciously, Jacob takes self-reproduction as the essence of the living but, above all, it is possible to recognize in this definition of the essence of the living the sedimentation of the oldest metaphysical matrix able to determine philosophies and the life sciences over the centuries: Aristotelian philosophy and, in particular, the teleology that, for Aristotle, consists in the essential structure of nature in general and of the living in particular:

> We say "this is for the sake of that" whenever there appears to be some end towards which the change proceeds if nothing impedes it. So it is apparent that there is something of this sort, which is precisely what we call a nature. Surely it is not any chance thing that comes to be from each seed, nor a chance seed which comes from a chance body; rather, *this* one comes from *that* one. Therefore the seed is an origin and is productive of what comes from it. For these things are by nature; at least they grow from seed. But prior even to this is what the seed is the seed of; for while the seed is becoming, the end is being. And prior again to both of these is what the seed is from. For the seed is a seed in two ways, *from* which and *of* which; that is, it is a seed both of what it come from, and it is a seed of what will be from it, though not in the same way, but of each in the way mentioned. Further, the seed is in potentiality; and we know how potentiality is related to complete actuality.⁵³

> For things posterior in generation are prior in nature, and the final stage in generation is primary in nature . . . For every generated thing develops from something and into something, that is, from an origin to an origin, from the primary mover that already has a certain nature to a certain shape or other such end. For a human being generates a human being, and a plant a plant, from the underlying matter of each. So the matter and the generation are necessarily prior in time, but in account the substantial being and the shape of each thing.⁵⁴

Derrida refers neither to this passage nor to other Aristotelian texts, but the allusion to this conceptual framework is evident:

> The fact that this internal property of the living (the capacity of reproducing itself) is not a hidden virtue but a logic of integration, this does not prevent us at all from considering it an essence: not only *ousia* (mode of being, of being as such, beingness [*étantité*]), not only essence as causality (*aitia*: moving and final causality as Jacob himself says), but also essence as *energeia*. . . . In other words, not only does Jacob not break purely and simply with philosophical discourse on the essence but, with the essence of life as tendency and ability to reproduce, he comes back not only to essence but to the essentiality of essence, the origin and end of the essence as the dynamics and energy of being, what gives the potentiality and act of being, the maximum of being, and secures—from the inside, this is the essence of the essence, that is, of having its principle of being in itself and not in the accident come from outside—secures from the inside its own production, namely, its re-production. From this point of view, not only is it difficult saying that for Jacob there is no essence of life but, on the contrary, he seems to affirm, in a traditional fashion, that life is the essence, the capacity of producing-reproducing itself from the inside (internal property), that it is in this sense more essential than the nonliving it integrates within itself, in its living being.[55]

The reference to Aristotle is made explicit where Derrida quickly describes the philosophical tradition determined by the theologico-metaphysical matrix of the essence of life that constitutes the legacy on which Jacob's discourse seems to rest unconsciously:

> He fits these remarks back into a very classical discourse on essence, the one that in Aristotle accords *dynamis* with *energeia* (through efficient and final causes) or the discourse of Spinoza's *conatus*, or of Leibniz's *appetitus*: it is evidently the Hegelian discourse, as it articulates the whole logic of essence with the value of life (natural life and life of the spirit), that seems here to be the closest one.[56]

The Absolute Programme 83

The choice to address the question of life (and death) between biology and philosophy starting from Hegel is fully justified: it is the German philosopher who takes up the Aristotelian legacy in the modern age by translating the theologico-metaphysical matrix of our conception of life into the Christian and speculative element. This is a heavy legacy for a biology that wants to be emancipated from philosophy by merely dissociating teleology from its theologico-metaphysical matrix—the divine as the intentional cause of the finality of nature—and at the same time leave intact the definition of the essence of life that the teleology elaborated by the philosophical tradition necessarily presupposes. Life as the internal property, as the internal organizing principle, as the irreducible condition that regulates from within the reproduction of the living, which therefore would occur independently and autonomously with respect to alterity and exteriority in general, by relating to them only belatedly, and thus with respect to death or, at least, as we shall see, with respect to a certain determination of death.[57] At this point for Derrida there are no doubts: biology is still the unconscious heir of the theologico-metaphysical presuppositions that have historically determined its genesis and development. However, this does not mean that he aims to restore the primacy of philosophy but to highlight the necessity of a different attitude of biology before its own history and the legacy it must assume *volens nolens*. He aims to highlight the necessity for a rigorous critical vigilance with respect to tradition that biology would otherwise end up repeating in the illusion of being definitively liberated, or, more precisely, the necessity for a deconstructive attitude in the interest of the progress of the sciences of the living:

> For the moment I am attempting to bring out the relation that the discourse of the geneticist, of the modern biologist has to the philosophical tradition: misrecognized debt and dependence, denial, subordination to the constrains of a code, of a programme, precisely, of a calculating machinery from which he believes he is emancipated whereas he is reproducing its functioning, etc. In referring too quickly and cumulatively to Aristotle, Leibniz, Spinoza, Hegel, Nietzsche, I didn't mean to assimilate all these systems and in for a confused amalgam, as far as they are concerned, but I meant to mark very rapidly that the discourse of modern genetics broke less with those classical philosophemes than it seemed or claimed. And that the fact of failing to explain oneself with respect to those classical philosophemes was not

justified, and rather, on the contrary, risked blinding us to the repetition of a very powerful code, which is itself somewhere an effect of the logic of the living.[58]

Despite, or rather thanks to, the supposed break with the philosophical tradition, biological discourse risks repeating and thus confirming the metaphysical matrix that founds and organizes the definition of the essence of life across our tradition. In so doing, it risks limiting the groundbreaking potential implicit in its recent discoveries. And this is the case not only for the logic of essence to which Jacob recurs but also for what Jacob determines as the essence of the living and thus "reproduction," and not only because this determination conjures up the Hegelian legacy. For Derrida, Jacob uses the notion of "reproduction" without any critical vigilance. There is no exact determination of the sense of the word or of its function in the biological context that would justify a rigorously scientific approach. For Jacob, the sense of this notion must be clear and self-evident, and for this reason he does not deem it necessary to linger on its provenance, on the history of the semantic and conceptual sedimentations that overdetermine its possible uses and significations. In the absence of precise operatory distinctions, Jacob unconsciously inherits through the notion of reproduction its semantico-conceptual history and inscribes it into the biological definition of the essence of the living. Therefore, for Derrida, he cannot grasp the destructuring and deconstructive bearing of his own definition of the essence of the living.

Let me recall first that, for Derrida, when Jacob recurs to "reproduction," we should read "self-reproduction" ("*se reproduire*"): "For Jacob always says re-production where he is clearly describing a self-reproduction: there are non-living beings that re-produce without re-producing themselves and this flexion onto themselves, this auto-affection is an essential fold of the structure."[59]

In a moment I focus on the destructuring bearing that, for Derrida, is implicit in this auto-affective "flexion" that structures the reproduction of the living and of which Jacob seems to be unaware. In order to understand it, we must first approach Derrida's critique of the naive use of the notion of reproduction mobilized by Jacob, among others.

Essence: Producibility and Reproducibility of the Living

Jacob gives "reproduction" a meaning that carries an implicit and thus uncritically accepted reference to the notion of "production" from which the meaning of "reproduction" should derive—at least, in appearance, that is, according to a semantic appearance:

> Jacob's discourse—like that of an entire modernity—handles the concept of production and reproduction as if it were transparent, univocal, self-evident, as if there were also a clear distinction or opposition between producing and reproducing, reproducing and self-reproducing. Jacob never asks himself what this means, he never submits this concept or this word of production/reproduction to the slightest critical question. Yet this is the major, ultimate, operative concept of his whole discourse. The logic of the living, the structure of the living and therefore the essence of the living are determined as productivity (self-re-productivity). And not only does he take for granted the clarity of the re- and the *self* that apparently specify and yet pre-determine production [*le produire*]; not only does he take them for granted, but he also takes for granted the clarity of the meaning of *producing*.[60]

The notion of reproduction, which is crucial in Jacob's discourse, depends therefore on the use of "production" as it has come to be commonly used in "modernity," to the extent that—this is Derrida's hypothesis—it responds to an urgency that characterizes the discourse of modernity as such, namely, as a historial and not merely historical emergence:[61]

> The historial [*historiale*] urgency of this question is signaled by the fact that the notion of production everywhere comes to fill the voids of modern discourse. These voids are not deficits but, on their contours, they mark that it is no longer possible, in decisive moments, to make use of expired values, which are out of date, and they are regularly replaced with production, so that this notion becomes the general stand-in for the determination of being. Where it is no longer possible to say creating (because it is God alone who is said to create and we are done with the theological), we say producing; where it is no longer possible to say generating, expressing, thinking, etc. where a concept rightly looks suspect because it imports too much from metaphysics, theology or ideology, we call on producing to replace or neutralize it. You know that today we do not form a system, a theory or a concept, we do not conceive of a concept, we do

not express something, but we produce a statement, we produce a theory, we produce an effect.[62]

The modern use of the concept of "production," the diffusion of which, Derrida suggests, should be retraced back to Marxism and to relative cultural hegemony, would be useful to express the modality of the determination of the being of beings by simultaneously liberating this determination from the theologico-metaphysical legacy implicit in analogous notions that belong to our tradition. However, given that, for Derrida, metaphysical concepts do not exist by themselves but rather in the order of metaphysical discourse that presides over the systematic articulation of concepts and their textual functioning,[63] we may wonder if, in order to set ourselves free from the order of the metaphysical discourse, it is enough to replace the exploited concepts of the metaphysical tradition with concepts that are allegedly "new" or "alien" with respect to such a tradition:

> I do not say this to provoke an effect of derision. The opposite is true: convinced of the historical necessity of this filtering and selection, first carried out to eliminate a whole set of values implied in the notions thus excluded or replaced, I am wondering what this vicariousness means. Through the selection and filtering thus carried out, a whole set of values (acting, creating, generating, thinking, etc. with their whole system, which is enormous) is marked as non-pertinent, excluded, except producing. What are we hoping to guard and re-produce here?[64]

Therefore, we must verify: 1) what we want to reject and what we want to preserve of the determination of being through the determination of the latter as production; 2) if this semantico-conceptual selection is sufficient to produce a break with the theologico-metaphysical tradition or if it rather entails the unconscious reproduction of the latter's value. For Derrida, despite the work of conceptual selection and substitution, the notion of production does not by itself represent a safe alternative to the order of metaphysical discourse. Conversely, a necessary reference to the fundamental value that structures and presides over that order, namely, presence, is inherent in it:

> Once again, these are not the questions of a philologist, nor even of a philosopher (since there are philosophemes that are

not problematized by philosophy and more than by science), but questions about the functioning of a certain number of discourses—and even discourses that at least within a certain field, are dominant—discourses that, however scientific they are up to a point (and up to this point their scientificity is not in question), nonetheless need an uncriticized, an unquestioned operatory support—here the notion of production—to ground their scientificity. This support is evidently a philosopheme (determination of being as *physis*, *techne-aletheia*-life, manifestation-information, production of essentiality as a maximal self-reproduction, self-presentation, etc. as what does producing mean but presenting? . . . This philosopheme that at a given moment seizes the whole foundation, supports the selection necessary to the progress of science, the elimination of the nonscientific, etc. there is no doubt that this dominant philosopheme serves science, but it is also through the body of this philosopheme that all the non-critical operations are going to pass and along with them, in the same movement, all the impositions that are designated by the name of ideology.[65]

In particular, this philosophical scheme—production/reproduction—does not only imply the unconscious and uncritical assumption of the fundamental value of metaphysics, namely, presence, and thus of the whole system of hierarchized oppositions that rests on that value, but it also prevents us from grasping the destructuring and, indeed, deconstructive, effects that are implicit in the determination of the essence of the living as self-reproduction. For Derrida, the concept of "self-reproduction," qua essence of the living and principle of the logic of the living, should necessarily entail a critical revision of what we understand as "logic" and "concept," to the extent that the meaning and function of these notions are structurally solidary and dependent on the value of presence that founds and governs the tradition of the Western thought.

> The concept of reproduction is hardly conceivable. A fortiori the concept of the re-production of self, of auto-reproducing, of self-reproduction. Especially if we claim to recognize in it an origin and an essence, the origin and essence of the living, the internal property of the living. . . . It is not a logic or a concept among others, an example of logic or concept, and this for the

primary reason that it obliges us to transform the usual logic and concept of the concept, and this for the pre-primary reason that it is the pre-primordial reason, on the basis of which logic and the concept are produced in general.[66]

Therefore, for Derrida, once we admit that self-reproduction consists in the essence and pre-condition that structure the life of the living and thus also of the human being, it would be necessary to verify if and to what extent this irreducible condition—self-reproduction—affects the "production" of the human being and thus also the notions of "logic," "concept," "presence," and "production" itself. To understand this audacious hypothesis, we must first take into account the radical consequences implicit in the thesis that self-reproduction consists in the essence of the living and thus in the irreducible condition of its structural possibility:

> Self-reproducing presupposes that one already is. As Jacob observes in passing "only what exists reproduces itself. Selection does not operate among the possibles, but among the existents." Therefore, self-reproduction reproduces what (self) already exists [ce (se) qui existe déjà]. But here what already exists is the effect of a self-reproduction. Of another self, the same. However far as we go back we won't find a reproduction that does not reproduce a re-production. An absolute self-production produces a self that is a (living) self only to the extent that and in that this originary and living production produces itself—produces itself—[se produit-produit soi-même-] as reproducibility. The self of self-producing is already, in its identity, reproducibility, without which it has no identity. Self-identity or the identity of the self to itself [l'identité du soi ou de soi à soi] is a certain reproducibility.[67]

Self-reproduction as the irreducible condition of possibility of the living implies that, in order to be itself and thus be self-reproduction, the living must differ from itself so as to relate to itself and reproduce itself. But this means first that the possibility of reproducibility—self-repetition or iteration—is neither given nor constituted by itself, it cannot be realized nor can it present itself spontaneously and independently from reproduction, but rather consists in (an effect of) self-reproduction. This means that the identity—the being itself—of the living will never be of the order of mere presence but always already of the order of re-production, which presupposes the irreducible condition of differing from itself that makes self-relation and thus self-reproduction possible:

> But it is precisely here that the definition of the logic of the living or of the essence of life as self-reproducibility makes self-relation, the relation of the self to itself, the self as self-relation into the essential fold of the living, a fold according to which self-producing—in other words, the living (only the living can produce itself)—would be self-reproducing. The *self* effaces as it were the difference between producing and reproducing. In self-reproducing neither the *self* nor the *re* come to affect from the outside, they do not supervene on a producing that would precede them, a product that would pre-exist them. What seems to pre-exist is already a re-produced as re-produced of the self, a self-reproduced. And when Jacob says "only what exists reproduces itself," what exists is already a produced as the effect of a self-reproduced. Therefore, self-re-production must be thought otherwise than what comes after the fact to complicate a simple production. Producibility is from the beginning re-producibility, and re-producibility is self-re-producibility. But, as the self is not before this capacity of self-reproducing, before its own re-producibility, it has no self-sufficiency nor pure spontaneity before production *as* reproduction, before its reproducibility as re-producibility. It is its reproducibility. Given which, production, that is, the producing of production is neither conceivable nor possible before re-producibility as self-re-producibility.[68]

At this point, it should be evident that, through Jacob's definition of self-reproduction as the essence of the living, that is, as the irreducible condition of the living, Derrida retrieves the apparently enigmatic definition of différance as the structural condition of life and evolution as it had been developed in "Freud and the Scene of Writing" in relation to arche-writing. Let me remark that this notion was elaborated in order to account for iterability, that is, repetition, as the irreducible condition of the constitution of the retentional trace and thus of memory, or, more precisely, of the psychic system, beyond the threshold of the constituted consciousness and of the difference between animal and human being:

> No doubt life protects itself by repetition, trace, différance (deferral). But we must be wary of this formulation: there is no life present *at first* which would then come to protect, postpone, or reserve itself in différance. The latter constitutes the essence of life. Or rather: as différance is not an essence, as it is not anything, *it is not* life, if Being is determined as *ousia*, presence, essence/existence, substance or subject. Life must be thought of as trace before Being may be determined as presence. This is the only condition on which we can say that life *is* death, that repetition and the beyond of the pleasure principle are native and congenital to that which they transgress.[69]

It is therefore on the basis of the essence of the living as self-reproduction that Derrida can extend the law of différance to the living in general, a law that allows the living to be structured in view of an arche-writing that is already at work at the level of its most elementary organization and of which psychic arche-writing would be another specific articulation, as well as other specific articulations would be its "products," that is, the products of the human being, such as the idealities, which consist in the instruments adopted for the knowledge of the living:

> If all the productions of the living—what we call the productions of the living and in particular the productions of the living being called man (culture, institution, *techne*, science, biology, the texts in the strict sense) have somewhere as a condition the production of the living as self-reproduction, and if, on the other hand, the supposed "models" required to understand or know the living are always themselves products or productions of the living, you see not only the twist of this logic but also the urgency of asking "what about production and self-re-production?"[70]

Thanks to Jacob's definition, Derrida opens up the way that will lead him to conceive of the living as a text and, at the same time, to deconstruct Jacob's recourse to writing as a mere analogical and operatory model. In other words, to understand the relationship between the text and the living as the articulation of the general textuality elaborated in "Freud and the Scene of Writing" and in *Of Grammatology*.

The Essence and the Supplement: Sex and Death

Despite or, perhaps, thanks to its supposed break with the philosophical tradition, biological discourse risks repeating and thus confirming the metaphysical matrix that has grounded and organized the definition of the essence of life across our tradition; it risks limiting the groundbreaking potential implicit in its discoveries. We may find another confirmation of the hypothesis advanced by Derrida if we consider that the definition of the essence of life derived from Jacob's discourse shows not only traits analogous to those elaborated across our tradition on the basis of a shared metaphysical matrix, but also the same consequences. For Jacob, sexuality and death are accidental, nonessential, even superfluous factors with regard to the essence of life understood as auto- or self-reproduction. Life in itself, in its essence, would not necessarily imply sexuality and death; it would recur to these "auxiliary" factors only when coming to a certain degree of complexity. Indeed, at the level of the simplest form of life—bacteria, that, from the point of view of evolutionary theory, are also the first form of life that appeared on earth and from which, therefore, other and more complex forms would have evolved—reproduction occurs through the fission of individuals and not the combination of the genetic material of the members of a couple:

> Evolution has become possible, only because genetic systems have themselves evolved. As organisms become more complicated, their reproduction also becomes more complicated. A whole series of mechanisms appears, always based on chance, which help to reassort the programmes and compel them to change. . . . But the most important inventions are sex and death. Sexuality seems to have arisen early in evolution. At first it was a kind of auxiliary of reproduction, a superfluous gadget, so to speak: nothing obliges a bacterium to make use of sexuality in order to multiply. It is the necessity of resorting to sex as a reproductive device that radically transforms the genetic system and the possibilities of variations. As soon as sexuality becomes obligatory, each genetic programme is no longer formed by exactly copying a single programme, but by reassorting two different programmes.[71]

In particular, as it occurs in Hegel—I refer to the conclusion of the *Philosophy of Nature*—Jacob claims that death is a consequence of sexual reproduction and thus would not affect life as such. As they reproduce by fission, for Jacob, bacteria do not die, just dilute:

> With bacteria, unlike organisms which reproduce only sexually, birth is not counterbalanced by death. When bacterial cultures grow, the individual bacteria do not die. They disappear as individuals: where there was only one, suddenly there are two. The molecules of the "mother" are distributed equally among her "daughters." For instance, the mother contained a long duplex of deoxyribonucleic acid that splits into two before cell division. Each daughter receives one of these identical duplexes, each of which is formed by an "old" chain and a "new" one. One of the criteria that a bacterium is no longer alive is its inability to reproduce. If this non-life is to be seen as death, it is a contingent death. It often depends on the conditions of the culture. When a small part of a culture is continuously replaced by fresh medium, such a culture remains in a state of perpetual growth: bacteria go on reproducing indefinitely [*éternéllement*]. What makes an individual ephemeral in a bacterial population is not, therefore, death in the usual meaning of the word, but dilution entailed by growth and multiplication.[72]

Derrida engages in a close reading of this decisive passage, in which he sees the unavoidable consequences of the reproduction of the metaphysical matrix of the essence of life: if auto-reproduction is the essence of life as an interior, autonomous, and independent property with respect to alterity and exteriority in general, as its own cause and end, then death, the other of life, is not only inessential, accidental, and contingent with respect to auto-reproduction, but also must be exterior to life, something that affects life from outside without affecting its essential property. The opposition inside/outside, also recognized as both the conceptual matrix and blind spot of metaphysics since "Plato's Pharmacy,"[73] is the ultimate target of the deconstruction of these passages:

> What this description leads Jacob to is to posit that in the system of the simple reproduction of the bacterium neither sexuality nor, *consequently*, death are essential constituents; and thus,

consequently, they happen as supplement, as from outside. It is on this link between sexuality and death, on this value of the outside, that, on the contrary, I would like to insist a little.[74]

For Jacob, the death of bacteria would not be a proper death, as it would be caused from outside, from the environmental conditions in which a population of bacteria grows or is put to grow. In optimal environmental conditions, the reproduction of bacteria and thus the life of bacteria can perpetuate "eternally"—thereby life, at the origin of evolution, would be without death:

> What, then, Jacob calls death in the habitual sense, that is, according to him, the only death that deserves this name, is a death that does not limit itself to a non-life, a death that would not be contingent, contingent translating here as "coming from outside," affecting from outside. Indeed, as you have seen, this contingency, which retains death within simple non-life, which prevents bacterium from having a right to death, to its own death, this contingency derives from the fact that death depends on the outside, on the milieu. . . . Let us reflect further upon what has just been put forth here. The death coming from outside (determined here as a milieu) is not death (in the sense proper determined on the basis of the "usual sense"). This non-life is a non-death, this non-life is not a death. Saying that the non-life that occurs to the bacterium is a non-death can be inverted into "life is death": true life is true death. I would say that this consequence is not absent from Jacob's text, as it will show later that death must be interior and essential to life in order to be truly death (a double consequence for bacterium and human being . . .). But consequence just as a logical can be also drawn from the same statement, namely that, as true death never reaches the bacterium, and as the latter is living in that it reproduces itself (this is the criterion), the life of the bacterium (a-sexual life multiplying itself through the simple division of the one), this life is invulnerable, pure life, inaccessible to the least negativity. Death does not affect it, it passes over it as its outside.[75]

Referring to the second "invention" that occurred in the history of the evolution of the living, namely, the invention of death, Jacob explains that death is a necessary and, thus, essential, internal property, determined by the genetic programme: "The other necessary condition for the very possibility of evolution is death. Not death from without, as the result of some accident; but death imposed from within, as a necessity prescribed from the egg onward by the genetic programme itself."[76] At this point, Derrida focuses on the paradoxical law that structures Jacob's discourse, the law of the supplement, as he puts it elsewhere. This law forces the discourse to recur to what it had determined earlier to be auxiliary, exterior, superfluous, in order to account for what is proper, interior, necessary—namely, the essence:

> And yet, here is the paradox of the graphic of the supplement, which Jacob does not take into account at all: it is that this death as internal prescription in living beings with sexual reproduction, this non-supplementary death intervenes in the chain of non-sexual reproduction as supplement. . . . One would then have to admit that for sexuality as well as for death, that these two "inventions," coming from outside, quasi-accidentally, consist in bringing inside, in inscribing as an internal law the very thing that comes from outside.[77]

Here and everywhere, the logic of the supplement makes the oppositions on which the discourse is grounded and structured unbearable by leading the discourse itself to contradictions and inconsistencies of various kinds. Above all, the inside/outside opposition becomes problematic, which evidently regulates any other opposition, as the outside becomes necessary to the definition of the inside to which it must be opposed in order to preserve the integrity of the inside. The outside is found inside the inside, while it necessarily remains outside:

> What the supplement brings in from outside is an internal supplement, so that all the oppositions that Jacob handles with assurance (necessary/contingent, interior/exterior, organism/milieu, etc. and consequently non-sexuality/sexuality, life/non-life) go awry, which obliges him, even if he never reflects upon this law, either to formally contradictory enunciations, or to empirical approximations in which the conceptual edge of certain statements falls down, crumbles or becomes blunt.[78]

Thanks to the law of the supplement—before or beyond the opposition between the philosophical discourse and the scientific one, equally destabilized by this law—the discourse remains imbricated in the effort to reconstruct the logic of the opposition on which its conceptual hold and demonstrative effectiveness rest. For Derrida, in Jacob's discourse this effort turns into the attempt to isolate the terms of oppositions through the construction of ad hoc explanatory models (for instance, the analogy between genetic memory and cerebral memory), that is, models that safeguard the functioning of oppositional logic and are at the same time abstract and unable to account for the contradictions to which the discourse necessarily falls prey:

> One witnesses then an effort (a properly philosophical effort) of the scientist, . . . a philosophical effort, then, to reconstitute the conceptual oppositions or the essentialities there where the logic of opposition (dialectical or not) or the logic of essence loses its pertinence. This effort always attempts to isolate or purify models (that are therefore pure) that would permit a secure handling of the binary or dialectical logic, i.e., the mastery of certain programmes that would be ultimately inaccessible to the supplement or in which the supplement itself would be incorporated into the programme.[79]

This effort ends up propagating the paradoxical effects of the logic of the supplement and leading the discourse through a series of unbearable contradictions and inconsistencies, as occurs in the case of bacterial reproduction by fission, which Jacob takes up as the model of non-sexual reproduction and places at the origin of the evolution of life. Derrida's critique of the opposition between non-sexual and sexual reproduction, as it is deconstructed in Jacob's discourse, calls into question not only Jacob's thesis that reproduction by fission would be the reproductive modality proper of bacteria and, thus, the model on the basis of which reproduction is determined as self-reproduction, but also, more radically, the idea that bacterial reproduction is alien to sexual reproduction. Jacob admits that bacterial reproduction can also occur according to other modalities analogous to sexual reproduction, and yet his effort consists in preserving the purity of the model and denying the title of sexual reproduction to these modalities:

Some bacteria also have another way of adding to their genetic programme. Most frequently, micro-organisms are isolated from each other. They do not communicate. They exchange nothing. They are even protected from any relationship by their cell wall. Nevertheless, transfers of genetic material from one cell to another sometimes occur, either by the intermediary of a virus, or by processes recalling the sexuality of higher organisms. But this addition of genetic material has a lasting effect on the descendants of the cell only in so far as the fragments thus introduced succeed in taking root there, reproducing and being transmitted from one generation to the next. Such an implantation often occurs by genetic recombination. So a segment of chromosome can be replaced by an homologous segment from another individual. Among populations of bacteria that multiply under different conditions, different genetic sets tend to be formed according to environmental requirements. Through recombination, the elements of genetic texts, genes from different individuals, can be reassorted in new combinations that sometimes offer advantages for reproduction. Even though sexuality is not really [*véritablement*] a method of reproduction for bacteria, that usually multiply by fission, it nevertheless allows the different genetic programmes of the species to be mixed with the resultant appearance of new genetic types.[80]

For Derrida, the mere possibility of forms of bacterial reproduction that are different from reproduction by fission is enough to undermine the model status given to the latter and, at the same time, betrays the interest behind the construction of an *ad hoc* model: the originary essence of life as pure auto-reproduction, independent from alterity in general and exteriority in particular:

Whatever the frequency or rarity of these phenomena, their partial character, they signify at any rate by their mere possibility, that this can always occur to the "pure" bacterium as model of reproduction without sexuality and without death, pure inside or pure outside, pure inside of living reproducibility or pure surface that can receive death only from outside, as contingent. From this I do not want to conclude that there has always been sexuality or death or, by simple inversion, that there will never yet have been either, but that if "science" or "philosophy" must speak of sexuality or death, the oppositions (positive/negative, more/less, inside/outside), the logic of "either/or," of the "and" and of the "is" is no longer sufficient for this. And I want to say that the concept of model is always there to mask that it is no longer sufficient.[81]

First, Jacob's distinction between para-sexual bacterial reproduction and *truly* [*véritablement*] sexual reproduction is not rigorous and cannot be sustained as such, precisely because of the definition of sexual reproduction given by Jacob himself: "As soon as sexuality becomes obligatory, each genetic programme is no longer formed by exactly copying a single programme, but by reassorting two different programmes."[82] If what identifies sexual reproduction is the combination of two different programmes or genetic texts, then there is no reason why this form of bacterial reproduction should be defined as non-sexual. Following Jacob's criterion, what is at stake between modalities of reproduction that are essentially different and opposed is not a simple analogy but, once again, a difference of degree within the same order, the order of the reproduction obtained through the combination of two different genetic programmes. Secondly, Derrida questions the very idea of non-sexual reproduction or, rather, the definition of reproduction by fission as auto-reproduction (1:1), as the exact copy of a single programme. This definition implies the rigorous closure of the bacterial cell, which would be immune to possible combinatorial interactions with other programmes or genetic texts and thus able to reproduce an exact copy of itself. Indeed, Jacob himself cannot rigorously exclude the possibility that a "pure" bacterial cell receives an additional segment of genetic text from outside; or rather, he admits that the phenomenon is a possible condition of sexual differentiation in bacteria:

> Recombination only reassorts the genetic programmes in populations; it does not add to them. Certain genetic elements are, however, transmitted from cell to cell and simply added to the genetic material already present. The instructions they contain are indispensable neither for growth nor for reproduction. But this addition to the genetic text allows the cell to acquire new structures and perform new functions. It is an element of this type that determines sexual differentiation in certain species of bacteria, for instance. Furthermore, as it is not indispensable, the

nucleic-acid sequence contained in such supernumerary elements is not subject to the constraints of stability that natural selection exercises on the bacterial chromosome. These elements represent a free addition for the cell, a sort of reserve of nucleic-acid text that can vary freely in the course of generation.[83]

Once again, Derrida finds the logic of the supplement and its unavoidable contradictory effects at work: if we cannot rigorously exclude the possibility of the addition of a supplement of genetic text to a bacterial cell, then we cannot rigorously affirm the pure identity closed upon itself. If an addition is always possible, then the opening must be a structural condition of the bacterial cell: the outside is already inside of the inside of which it would be a mere supplement. Hence, what biology considers supplements outside of life—sexuality and death—are rather conditions of possibility of life itself.

Also in this case, the most recent achievements of research seem to confirm Derrida's argument: Hema Prasad Narra and Howard Ochman, in "Of What Use Is Sex to Bacteria?," claim that the reproduction of bacteria through the assimilation of a genetic material (DNA) that is different from that of the "mother" cell must be designated as a sexual reproduction:

> Despite their asexual mode of reproduction, bacteria have sex, or at least something that is referred to as sex and can be defined as the inheritance of DNA from any source aside from the parental cell. Unlike the sexual process occurring in most eukaryotes, the transfer of genetic material during bacterial sex is unidirectional and can occur by one of three mechanisms that differ with respect to the source of DNA and/or the types of partners involved. Although these processes—conjugation, transformation and transduction—were each characterized decades ago, the availability of complete bacterial genome sequences has brought renewed interest in their contributions to the contents and organization of bacterial genomes, their consequences on attempts to reconstruct the phylogeny and relationships among bacteria, and their role in bacterial adaptation and the dissemination of disease determinants.[84]

Furthermore, Narra and Ocham demonstrate that the mechanisms of genetic transfer that can be observed in the reproduction of bacteria, even if they have been known for a long time—also to Jacob, as we have seen—appear as anything but occasional and marginal. Indeed, they are quite diffused and play an essential role in the evolution of bacteria, to the extent that they introduce a greater genetic variety and thus secure better conditions of survival than the reproduction 1:1.[85] In particular, the bacterium *Escherichia coli*, which, at the time of Jacob, constitutes the experimental base for the thesis of the asexual reproduction of bacteria, presents in its genetic composition the effects of a particular type of sexual reproduction: the "lateral gene transfer," that is, the introduction of a completely novel sequences in the genome. Narra and Ocham thus conclude "that the majority of genes in all bacterial genomes were acquired laterally at some time during the evolutionary history of the lineage."[86]

If the sexual reproduction of bacteria is not an occasional and marginal phenomenon, but an essential condition of their evolution, a neat opposition between asexual and sexual reproduction in the life of bacteria can no longer be secured. The evolutionary necessity of sexual reproduction is inscribed in the evolutionary limits of asexual reproduction, according to the logic of the supplement highlighted by Derrida:

> While it is easy to point to the potential benefits of sex and it is known that gene transfer has affected the contents of virtually all bacterial genomes, the origin of sex in asexually reproducing lineages is less clear. It has been argued that each of the mechanisms that currently effect the transfer and uptake of DNA originally served purposes other than sex. It is certainly plausible that transducing bacteriophages might erroneously package and transmit host genes during their normal parasitic life cycle; however, the processes of transformation and conjugation seem to have evolved to transfer and obtain DNA from outside sources as requisite for sex.[87]

However, for Derrida, one could push the logic of the supplement to its limits, up to the most elementary conditions of reproduction within a single cell.

For Jacob, before the reproduction of complete living forms, and thus within the cell, there is only a chemical component that is able to reproduce itself as the copy of itself, namely, the deoxyribonucleic acid of the chromosomes that composes the genetic programme presiding over the reproduction of the cell and thus the mechanism of heredity.[88] Thanks to the complementary double helix structure, the deoxyribonucleic acid works as a text and makes the reproducibility of the genetic programme, its own copy, possible:

> Deoxyribonucleic acid is, in fact, a long polymer formed not of one, but of two chains, helically twisted around each other. Each chain contains a skeleton formed of alternating sugar and phosphate groups. Each sugar molecule is linked to only one chemical residue—an organic base—of which there are four different kinds. These four sub-units are repeated by millions in infinitely varied combinations and permutations along the chain. By analogy, this linear sequence is often compared to the arrangement of the letters of the alphabet in a text. Whether in a book or a chromosome, the specificity comes from the order in which the sub-units, letters or organic bases, are arranged. But what gives this polymer a unique role in reproduction is the nature of the relations that unite the two chains. Each organic base in one chain is associated with one in the other, but not just any one. The system of chemical bonding is such that each sub-unit on one chain can correspond to only one of the other three sub-units in the second chain. If the four sub-units are indicated by A, B, C and D, A in one chain is always opposite B in the other, and D is always opposite C. The symbols go in pairs; the two chains are complementary. The sequence in one chain imposes the sequence in the other. Owing to these peculiarities of structure, the nucleic acid duplex is exactly reproduced.[89]

Finding in this passage the opening of the analogy between "genetic programme" and "text," Derrida highlights the stakes of this passage. If reproduction by 1:1 duplication is made possible by the dual structure of the acid, which is differentiated in itself, then we could affirm (a) that 1:1 reproduction, through the fission of the bacterium, is already an effect of the reproduction of two terms within the cell, and thus this duplicity is the irreducible condition of reproduction; (b) that it is precisely this differential structure that determines the structural opening, the self-differing, of the cell attested to by the contradictions produced by the logic of the supplement. Derrida's reading follows this direction:

Proteins do not reproduce themselves. They depend on their organization and on the reproduction of something else that, instead, reproduces itself, spontaneously, by producing a copy of itself, and this is the deoxyribonucleic acid. According to Jacob, this is the only element in the cell that can reproduce itself by reproducing "copies" of itself. This power of producing copies of itself depends on the fact that the acid is formed by two chains, each double (sugar and phosphate) and it is due to this duplicity that permits the internal that the first textualization is produced as reproducibility. How does this duplicity relate to the supplementarity we have just spoken about?[90]

To answer this question, Derrida must finally examine Jacob's recourse to the notion of "writing" and thus the analogy or, rather, the consequent relationship between the text and the living.

IV

The Text and the Living

In the seminar *La vie la mort*, Derrida aims to verify the deconstructive effects that follow from the recourse to the text as the analogical model that would make it possible to explain the logic of the living, that is, the mechanism of its self-reproduction, which for Jacob amounts to the very essence of the living. First, what kind of text plays the role of analogical model: are we before a notion of traditional text or, rather, of a deconstructive one? Second, what are the consequences that derive from the fact that the living can be interpreted on the basis of a deconstructive textuality? As we saw, the theory of organized systems, developed by Wiener in the field of cybernetics, represents the reference model for molecular biology: it allows the latter to interpret the organization of the living in terms of programme, code, message, and communication. However, when it is necessary to explain the role of DNA in cellular reproduction, Jacob constantly recurs to textual models, from the Morse alphabet to the book:

> The model that best describes our knowledge of heredity is indeed that of a chemical message. Not a message written in ideograms like Chinese, but with an alphabet like that of the *morse* code. Just as a sentence represents a segment of text, so a gene corresponds to a segment of nucleic acid. In both cases, an isolated symbol means nothing; only a combination of symbols has any "sense." In both cases, a given sequence, sentence or gene, begins and ends with special "punctuation" marks. The

transformation of a nucleic-acid sequence into a protein sequence is like the translation of a message received in *morse* that does not make sense until it is translated, into English, for example. This is done by means of a "code" that provides the equivalence of signs between the two "alphabets."[1]

The DNA, which presides over the logic of the living, functions as a text: the sense of the genetic message does not depend on its alleged content but on the order of combination of the elements, which produces a sequence of interactions in the cell. Not every element produces effects by itself, just as the letter of the alphabet has no meaning in itself. For obvious reasons, Derrida emphasizes the preeminence of the syntactic order above the semantic one in the elaboration of the living text.[2] However, those traits are not enough to establish that the text employed by biology as an analogical model of the living is of a deconstructive kind. The opposite is true: so long as the text is a means for the transmission of the message codified by a programme, it is interpreted according to rather traditional semio-linguistic categories. The following passage from *The Logic of Life* is a good example: "A genetic code is like a language: even if they are only due to chance, once the relations between 'sign' and 'meaning' are established, they cannot be changed."[3] According to these coordinates, the recourse to the text as a model for interpreting the logic of the living imports, confirms, and renews the oldest metaphysical presuppositions on which the philosophical interpretation of life from Aristotle onward has been grounded. I recall what Canguilhem argues in the article quoted by Derrida:

> When we say that biological heredity is the communication of a certain kind of information, we hark back in a way to the Aristotelian philosophy with which we began. . . . To say that heredity is the communication of information is, in a sense, to acknowledge that there is a *logos* inscribed preserved and transmitted in living things. Life has always done—without writing, long before writing even existed—what humans have sought to do with engraving, writing and printing, namely, to transmit messages. The science of life no longer resembles a portrait of life, as it could when it consisted in the description and classification of species. . . . But it does resemble grammar, semantics and the theory of syntax.[4]

However, Jacob himself provides Derrida with the textual lever that allows him to deconstruct the text employed as an analogical model. He breaches the way for a deconstructive interpretation of the living that will have disruptive effects on the logic of the living. Apropos of the translation of the genetic message retained by the DNA in the proteic sequence, Jacob writes: "The genetic message can be translated only by the products of its own proper translation. Without nucleic acids, proteins have no future. Without proteins, nucleic acids remain inert."[5] Hence, for Derrida, the sense of the message is not constituted by itself, independently from its transmission/translation, according to the classical relation signifier/signified. The sense of the message consists in the effect that produces its transmission/translation; the sense therefore refers to another, different text; it necessarily and irreducibly bears within itself its differing from itself, according to the structural and temporal dynamic of différance and thus is never accomplished and established by itself, beyond the network of references in which it is inscribed. Therefore, Jacob's recourse to the text as a model for describing the logic of the living is not accidental. The text has imposed itself because of the very nature of the living, of the structure of the genetic message, the registration and the transmission, which renders reproduction and, thus, life possible: "When the first event, the real origin, etc., is a text, has the structure of a text, this fabulous adventure can always reproduce itself. This is what happens with the living being if it has the structure of a text."[6] Moreover, for Derrida, the necessity of this textual structure allows us to account for the logic of the living in more rigorous terms than those imported from cybernetics, and, therefore, to set biology free from the metaphysical remains that still resist within the naive use of notions such as "language," "code," "message," and "communication":

> There are of course effects of message, information and communication, but on condition that they be in the final analysis textual, i.e. that the message, the communication, the information never transmit (never emit, communicate, inform) any content that not itself be of the order of message, information, communication, thus not be itself a trace or a gramme. . . . The message emits a message: that appears to be a tautology but it is in fact the contrary of something obvious to common sense. The message does not emit something, it says nothing, it communicates nothing: what it emits has the same structure as it,

i.e. it is a message, and it is this emitted message that is going to allow the decipherment or translation of the emitting message, which implies the absence of anything outside the message, the information, the communication. This is why we have to be clear here that the words communication, information, message, are intra-textual and operate on condition of text, contrary to what they ordinarily lead one to think, namely that they communicate, emit or inform of something. Naturally this textual self-reference, this closure onto itself of the text that refers only to text has nothing tautological or autistic about it. On the contrary. It is because alterity is irreducible here that there is only text, it is because no term, no element here has any self-sufficiency nor even any effect that does not refer to the other and never to itself that there is text; and it is because the so called text cannot close onto itself that there is only text, and that the so called "general" text (an obviously dangerous and merely polemical expression) is neither a set nor a totality: it can neither comprehend itself nor be comprehended. But it can be written and read, which is something else.[7]

We should think the differential iteration of the self as the origin and sense of the life of the living and, thus, the possibility of the iterable trace as its irreducible condition. We should admit that the living is structured as a system of arche-writing, as the retention, elaboration, and protention of a weave, a tissue of traces, namely, as a text. Then the logic of the living is the logic of the "general text" elaborated in *Of Grammatology*, and it is only from this perspective that we can clear the notion of "general text" of the misunderstandings originated from the famous thesis "There is no outside-the-text [*il n'y a pas d'hors-texte*]."[8] This is not a thesis formulated to support the ontological radicalization of hermeneutics. This does not at all mean, in the wake of Heidegger and Gadamer, that only what is written in a book is the being that can be understood and, above all, that we can interpret everything as we interpret a book. Rather, it says that the trace is the condition of possibility of the living as well as of the non-living: as a condition of possibility of the living, it structures the life of the living as a text, from the biological organization to the texts written by humans living in their cultural habitat, passing through the animal arche-writing—as we saw, this evolutionary passage is decisive—which determines the relation of the living to the environment by responding to the laws of survival. Therefore, the condition is always the same, but at every step the elaboration of the text is different, the programme and the code of decipherment evolve in complexity and flexibility toward the limited freedom of man, in order to respond to the necessities of survival, which, in turn, significantly vary from the cell to social life, to the sphere of scientific knowledge:

> What might have seemed, more or less naively, to be the limited condition of philology, of literary criticism, of the science of documents and archives, etc., namely that of having as its ultimate referent something that we called, that we thought we knew, by the name "text," this condition is now that of genetics or of life-science in general; and if life-science is not one science among others, but also the science implied by all the sciences that determine their object in the fields that involve the living being (psychoanalysis, history, sociology)—all the human sciences but also all sciences insofar as they involve the activity of a living being—so all the sciences, all discourses and productions in general; if then life-science is not one science among others, then its textualisation, the textualisation of its object and of its subject leaves nothing outside it. This obviously does not lead, as might be claimed with a more or less interested or interesting naivety, that everything will be reduced, through the effect of this textualisation, to the snug inside of a book, a notepad or a more or less specialized library, but on the contrary leads to a very violent reinterpretation of the limit between this inside and its outside.[9]

However, looking closely, even in the edited texts, Derrida had always been explicit about the correct interpretation of the formula "there is no outside-the-text" and more generally about his extensive conception of the text as the implication of the dynamic of différence. In particular, in "Living On," which was written a little later than *La vie la mort* and, as we shall see, can be read as a consequential development of the seminar. It is not by chance, I believe, that Derrida explicitly alludes there to life as a text, as a weave of differential traces:

> What has happened, if it has happened, is a sort of overrun [*débordement*] that spoils all these boundaries and divisions and forces us to extend the accredited concept, the dominant notion

of a "text," of what I still call a "text" for strategic reasons, in part—a "text" that is henceforth no longer a finished corpus of writing, some content enclosed in a book or its margins, but a differential network, a factory of traces referring endlessly to something other than itself, to other differential traces. Thus the text overruns all the limits assigned to it so far (not submerging or drowning them in an undifferentiated homogeneity, but rather making them more complex, dividing and multiplying strokes and lines)—all the limits, everything that was to be set up in opposition to writing (speech, life, the world, the real, history, and what not, every field of reference—to body or mind, conscious or unconscious, politics, economics, and so forth).[10]

The implications of these arguments seem to be incalculable for the interpretation of Derrida's work on life and death and, more generally, of the very sense of deconstruction. The structure of the living is irreducibly open onto alterity in general, through which it relates to itself in order to be what it is and thus to reproduce itself. However, if the living must differ from itself in order to be itself, and if this difference is irreducible, then auto-reproduction will never be absolute, accomplished in itself, but only relative and open onto the difference from itself, onto alterity in general. According to Derrida, this is what Jacob too must admit when he speaks of the life of the cell that depends on energetic interactions with the other cells in the biotic environment in which it is immersed:

[I]s there not, also recognized by Jacob, a structural outside of the cell, without which the cell would not reproduce itself and that therefore makes the *self*, the self-relation of reproduction an always fissured and open structure, a system that works only when related to the other or the outside, so that the identity of the self- and of the re- is and functions only in its difference from itself, in the living as well as in the non-living.[11]

If, as we shall see, this is the germ from which Derrida develops his interpretation of life in terms of survival, then I take into account here some epistemological implications. If we replace the text understood in a traditional sense with general textuality, then it does not make sense to speak of a model. The text can play the role of analogical model for the living, it can be compared to the living, only if the text and the living are entities or autonomous terms that are constituted in themselves. Yet if the opening or the referral to alterity is the irreducible condition of both in their essential differing from themselves, then it is not possible to establish their reciprocal exteriority and independence, nor is it possible to posit them as such, nor, evidently, to relate the one to the other as analogical terms. Hence, on the one hand, it is no longer possible to speak of a model; on the other, and this is a consequence, it is possible to explain otherwise the relation between the text and the living, in particular, the very fact that it has been possible to interpret that relation, even if naively, according to the logic of the model. First, given the textual structure shared by the text and the living, thanks to which both must refer to something other than themselves in order to be themselves, the text and the living are inscribed in a homogeneous order that necessarily implies the possibility of their differential relation. Second, it is possible to explain their relation and thus the textual structure shared by the living and the texts (as they are traditionally understood), insofar as the latter are products of the former because they satisfy the same conditions of structural possibility that Derrida describes in terms of general textuality. Therefore, the condition is always the same, but at every step the living arche-writing is itself inscribed in different and ever more complex environments, the elaboration of the text is different, as are the effects of translation that it produces.

> In this case, the text can no longer be a model, a determined model, something to which one can compare something else. If at least there is a model and an analogy here, in this case they cannot be a model and an analogy among others. This is due to the structure of the living and of the text that can no longer play between them the roles of the compared and the comparing. If the text in the narrow sense (let's call it, in a vulgar way, the text as human production) is somewhere a production of the living being, it cannot be a model to which one can compare the living being of which it is an effect. But no more can it be thus compared if we extend the concept of textuality to the point of making it coextensive with the living being. In this case, it becomes meaningless and useless to speak of model. We are dealing with a sort of synonymy, equivalence, or redundancy.[12]

At this point, once the textual structure of the life of the living is established, according to Derrida, it is possible, if not necessary, to deduce a general epistemological principle: if the living is a text that produces texts, then we can legitimate the possibility of scientific knowledge, whose conditions and structures, however, we should reconsider from a textual perspective:

> This situation—a text without external reference, all on the outside because it has no reference other than a text remarking a text—is this situation not ultimately that of the text of bio-genetics which is written on a text of which it forms a part or of which it is the product, which writes on an object or a referent which not only in its turn is already a text, but a text without which the scientific text—itself the product of a living being—could not be written. The scientific text is of course in the situation that Jacob describes about and in his object—namely the living cell: he is one of the translators who are to the genetic message like the product of its translation. The activity of the scientist, science, the text of genetic science as a whole are determined as products of their object, if you will, products of the life they are studying, textual products of the text they are translating or deciphering or whose procedures of deciphering they are deciphering. And this, which appears as a limit to objectivity, is also—by virtue of the structural law according to which a message can only be translated by the very products of its own translation—the condition of scientificity, in this domain, of the effectuation of science and all the sciences. It is on this condition that the translation or decipherment (a decipherment that is neither objective, in the classical sense of this term, nor subjective, nor a hermeneutic of meaning or an unveiling of truth) it is on this condition that intra-textual decipherment is possible in this textual science without extra-textual reference.[13]

What first follows from this passage is that the traditional opposition, still accepted today, between the sciences of spirit and those of nature, Humanities and Sciences, is no longer rigorously sustainable. Their separation is still a legacy of tradition, a metaphysical legacy, which represents an obstacle to their development:

> Therefore, how can we still oppose the science of nature to—to what?—to the science of culture, society, man, and spirit? If there is somewhere a homogeneity (differentiated but of the same type) between the productions of the living called man (texts, in the narrow sense, calculators, programmers, etc.) and

the functioning of genetic reproduction, the opposition between the sciences of nature and other sciences loses its pertinence and rigor, and one wonders if biology can still claim to construct *its* truth, a truth of its specific field.[14]

For Derrida, we can rethink the articulation between sciences of nature and sciences of the spirit on the basis of the general text that constitutes their common matrix. However, I note that this does not mean accepting whatever kind of reductionism in either direction, scientific or spiritualist. As I have said and repeated several times, in this passage the differential/differing structure of the general text implies that, at any stage of differentiation, arche-writing is channeled into structural and specific emergences, through the rearticulation of the conditions inscribed in the matrix. But there is also a consequence of epistemological order that is necessary to take into account to understand the deconstructive significance of the interpretation of the living as a text. The tropes of writing—analogies, metaphors, images, and so forth—through which it is possible to describe and explain the text as well as the living, the one in relation with the other, are not mere illustrations of pedagogical nature, external to so-called scientific discourse, but contribute to the construction of that discourse by orienting the latter and the sense given to the text and the living. For this reason, in concluding the part of the seminar dedicated to Jacob, Derrida takes up the question of the model as an operatory factor that structures biological research in general. In this case, we may recall the heuristic function that Jacob ascribes to the use of concrete models in biology: "in order to last, a biological theory had to be based on a concrete model."[15] Derrida aims to verify what sort of text Jacob uses as a model of the living and how he justifies this use. To this end, he will uncover the theoretical interests to which this determination of the text as the model of the living responds, that is, the conception of the living of which the adopted text serves as a model by making its determination possible:

> Why, then, should we speak of text? Well, I believe that this necessity obviously has nothing absolute about it, nothing that is not bound and motivated by a certain historico-theoretical situation and to the politico-scientific strategy related to this situation. By referring the living to the structure of a text, we evidently make conceptual progress in bio-genetics, a progress in knowledge, in the knowledge of the living, if you like, it being understood that this progress of knowledge is at the same time a transformation of the status of knowledge that, as I said

> last week, has no longer to do with some meta-textual real but with the text and thus consists in writing text on text. It is not the recourse to the textual "model" that has made progress possible, but just as much the opposite: a certain transformation of knowledge imposed what we call the model of the text. Conversely, what we call model allows for new hypotheses, new constructions, and is in turn determined by that of which it is the model: we understand otherwise a text, what a text is, when the function called model is at work. This is when, whatever the inadequation of the concept and the word "model," we become aware of the necessity of this theoretico-political strategy I was speaking about a moment ago.[16]

Finally, Derrida aims to account for the most stringent reasons that impose the recourse to the notion of "general textuality" against what is traditionally understood as text, which is adopted by Jacob, even if in its cybernetic variant, as the model of the living:

> There is not the living and the text. Not only there are typical structures of the living and typical structures of text but, even if one is not content with the empiricist positing of this multiplicity, there are several ways of defining the textuality and structure of the living. It is obvious that if we determine textuality this time on the basis of a model of text (for example, the phonetico-logocentric text, oriented by a present meaning, etc., etc.), we are immediately involved in a system of interpretation of the living that is different, or even, opposite to the one that would subordinate this type of text to another (non-phonocentric, non-teleological, etc. etc.). The question of the model is then displaced and becomes: which type of text will serve as model for the science of general textuality. Is there a model text for general textuality?[17]

If the model-text adopted by Jacob intimates and justifies a conception of life that is structurally consistent with the conceptions produced across the tradition of the philosophies of life and thus dependent on the metaphysical matrix, it is in view of a different (*différante*) conception of the life of the living that we should recur to the general text elaborated within the horizon of deconstruction. However, Derrida does not aim to contrast the model adopted by Jacob with another model. What is at stake is rather deconstructing the very notion of model to point out that the heuristic function ascribed to the cybernetic text as a model of the living is not only unable to account for the living itself but necessarily entails the reduction of what in the living escapes the hold of the cybernetic model while remaining the irreducible condition of the latter's structural possibility. As we shall see, the necessity of referring to general textuality in view of describing these conditions derives from the deconstruction of the model adopted by Jacob, not only of the cybernetic text but also of the value and function of the model in general. In this case, the inadequate formal rigor in the construction and use of the cybernetic model represents for Derrida the symptom of an incomplete repression and justifies the recourse to general textuality in order to explain the genesis and structure of the living and thus to unfold a conception of life (death) that is other than the traditional one.[18]

The Model and the Living

Derrida observes that, for Jacob, the model first plays a role that is essentially analogical, with all the related problems, as we saw apropos of the notion of "programme":

> It [the model] does appear in Jacob's book, but not very often, and in a chain of substitutions in which equivalents for it are found by analogy, image, comparison or syntactical constructs like "just as, just as." This means that the model he speaks of is always a descriptive model, which highlights a resemblance, a natural affinity, and not a mathematical model, a model in the mathematical sense of the word. . . . The models Jacob talks about are, therefore, concrete, intuitive, and descriptive models, and perceptions of resemblances.[19]

According to Jacob, the notion of information, imported from cybernetics, had imposed itself in biology as it accounts for the essential trait that demarcates the living from the non-living: the tendency to conserve the order of the system against the general tendency that characterizes the whole matter according to the second principle of thermodynamics, that is to say, the tendency to the irreversible dissipation of energy, to entropy, which necessarily entails a more general tendency to go from order to disorder.[20] Even if a living system cannot elude this general condition of matter, it is able to employ the information drawn from the environment in view of conserving an organized state and thus of delaying—deferring—the ineludible effects of entropy (death). The organized systems are structured and conserved thanks to the exchange of information within the system, which allows them to oppose the general tendency to disorder through the reorganization and adaptation of the system on the basis of the contingent conditions of exercise, in particular, by draining energy from outside in order to compensate the dissipation of energy within the system:

> The system only works by means of a series of successive transformations involving information. Entropy and information are as closely connected as the two sides of a coin. In any given system, entropy provides a measure of both the disorder and man's ignorance of the internal structure; and information of both the order and man's knowledge. Entropy and information are evaluated in the same way. One is the negative of the other. This isomorphism of entropy and information establishes a link between the two forms of power: the power to do and the power to direct what is done. In an organized system, whether living or not, the exchanges, not only of matter and energy, but also of information, unite the components. Information, an abstract entity, becomes the point of junction of the different types of order. It is at one and the same time what is measured, what is transmitted and what is transformed. Every interaction between the members of an organization can accordingly be considered as a problem of communication.[21]

For Jacob, the cybernetic model can be applied to the living because information is an abstract entity that structures all organized systems as systems of communication, independently from their nature, whether living or non-living, artificial or natural:

> This applies just as much to a human society as to a living organism or an automatic device. In each of these objects, cybernetics finds a model that can be applied to the others: a society, because language constitutes a typical system of interaction

between elements of an integrated whole; an organism, because homeostasis provides an example of all the phenomena working against the general trend towards disorder; an automatic device, because the way its circuits are geared defines the requirements of integration. In the end, any organized system can be analyzed by means of two concepts: message and feedback regulation.[22]

Carefully considered, the cybernetic model is no doubt an abstract model, but its abstraction, its construction as a model, hinges on the reference to other models that are not at all abstract and are in turn borrowed from various specific contexts: nature, society, technology. Here, evidently, the problems related to Jacob's use of the notion of the model emerge. For the moment, Derrida limits himself to speaking of a use that is quite relaxed, not rigorous: "We may wonder what the epistemological value, or even the heuristic value of the model, can be, once it serves as a model for an object that is just as much its own model."[23]

However, for Derrida, the stakes of these passages are much higher, even decisive, given the consequences that derive from the conception of the living that the model implies and produces: the exclusive reduction, if not the repression, of the energetic, economic dimension of the life of the living; the dissipation, acquisition, conservation, and exchange of energy that characterize the life of the living:

> You may have remarked through the argument that I have just recalled the surreptitious displacement that has just been operated: as information is inseparable from entropy, as there is an exchange of matter and energy as well as of information, there is only exchange. And since the concept of exchange is in the dominant position of the most general concept, one passes from exchange to communication and privileges exchange as exchange of information over the exchange of matter and energy. This is the way the circulation of the model works: once the exchange in relation to the exchanged content (matter, energy or information, that is, selection/discrimination/election) has been privileged, it is easy to privilege among the exchange contents this content without content that is information, as it first consists in a selection or discrimination. Then, one says that everything in a system is information and thus communication.[24]

And yet the determination of information already entails by itself an activity of selection, it is already motivated in itself by the necessity of acquiring and transmitting—exchanging—energy. Therefore, we can neither repress nor reduce the energetic, economic dimension of the organization of the living; conversely, the energetic dimension must be recognized as a condition of the determination of information:

> The fact remains that, if information, the emission or reception of messages is itself inseparable from an activity of sorting, selection, that is, of force or difference of force, etc., and if information is not merely communication, language or neutral knowledge, one will not be able to isolate a pure linguistic or semiotic model from a, let's say, dynamic, energetic or economic one. The circulating, circular model is both informational and energetic. What we might have wanted—and we always might want to—surreptitiously eliminate by privileging message or communication, or the form—that is, energetics—does not allow itself be reduced and does not content itself with being added or coupled with the message, but structures—for instance, as selection or principle of selection—the message, the informational activity itself. And once one speaks of textuality, the value of relations of force, of difference of force, economical agonistic will also be irreducible therein.[25]

Here it seems that Derrida aims to retrace Jacob's strategy, the surreptitious operation of reducing force in favor of form or signification, back to the most general strategy of structuralism, in the wake of the critical reading elaborated in "Force and Signification," in which, as we recall, he had found in structuralism a variant of "preformationism" and more generally of the metaphysics of presence: "Jacob writes, classically dissociating, let's say, form and force (information and genetic regulation)."[26] In "Force and Signification," Derrida defines structuralism as a fascination for form due to the inability to think of force:

> Since we take nourishment from the fecundity of structuralism, it is too soon to dispel our dream. We must muse upon what it might signify from within it. In the future it will be interpreted, perhaps, as a relaxation, if not a lapse, of the attention given to

force, which is the tension of force itself. Form fascinates when one no longer has the force to understand force from within itself. That is, to create.[27]

Structure is then the unity of a form and a meaning.[28]

Therefore, Jacob's model is a structuralist model. The same criticism, directed toward the legitimacy and heuristic relevance of the dissociation between force and signification, which is required by the construction of the structuralist model, should also work for the construction of Jacob's model, which privileges the structure over the energetic dynamic that conditions its construction: "Force is the other of language without which language would not be what it is."[29] In "Force and Signification," Derrida highlights the coordinates of the work of deconstruction that the opposition form (signification)/force should undergo, in view of the determination of their differential/differing relationship:

> Our intention here is not, through the simple motions of balancing, equilibration or overturning, to oppose duration to space, quality to quantity, force to form, the depth of meaning or value to the surface of figures. Quite to the contrary. To counter this simple alternative, to counter the simple choice of one of the terms or one of the series against the other, we maintain that it is necessary to seek new concepts and new models, an *economy* escaping this system of metaphysical oppositions. This economy would not be an energetics of pure, shapeless force. The differences examined *simultaneously* [*à la fois*] would be differences of site and differences of force. If we appear to oppose one series to the other, it is because from within the classical system we wish to make apparent the noncritical privilege naively granted to the other series by a certain structuralism. Our discourse irreducibly belongs to the system of metaphysical oppositions. The break with this structure of belonging can be announced only through a *certain* organization, a certain *strategic* arrangement which, within the field of metaphysical opposition, uses the strengths of the field to turn its own *stratagems* against it, producing a *force of dislocation* that spreads itself throughout the entire system, fissuring it in every direction and thoroughly *delimiting* it [le *dé-limitant* de part en part].[30]

It is from this perspective, or according to this programme, that Derrida deconstructs the opposition form/force that structures the distinction between "message" and "feedback regulation" and thus the subordination of the latter to the former. In fact, the definition of the notion of "message" given by Jacob implies the reference to an activity of selection that would be its condition of possibility and presupposes the impossibility of reducing the energetic, economic dimension to the merely semiotic and informational one. The necessity of recognizing, acquiring, and transmitting information depends on the necessity of recognizing, acquiring, and transmitting energy, and this activity of selection requires with equal necessity the employment of energy:

> Message means a series of symbols taken from a certain repertory signs, letters, sounds, phonemes, etc. A given message thus represents a particular selection among all the arrangements possible. It is a particular order among all those permitted by the combinative system of symbols. Information measures the freedom of choice, and thus the improbability of the message; but it is unaware of the semantic content. Any material structure can therefore be compared to a message, since the nature and position of its components, atoms or molecules, are the result of a choice made from a series of possible combinations.[31]

Therefore, the activity of selection—the choice—does not intervene only at the level of the feedback regulation that allows the living organism to oppose the general tendency to entropy through the acquisition and transmission of the energy required to its own conservation. Rather, the selective activity of the feedback regulation is possible only if it is involved in the determination of information in view of its transmission:

> Feedback is a principle of regulation that allows a machine to adjust its activity, not only in terms of what it has to do, but also in terms of what it actually is doing. It operates by introducing into the system the results of its past activity. This brings into action sense organs responsible for estimating the activity of the motor organs, for verifying their performances and making the necessary corrections. This supervision is meant to correct the mechanism's tendency towards disorganization, that is, to reverse temporarily and locally the direction of entropy. These

mechanisms range in complexity from the simple regulation of a boiler in relation to the surrounding temperature, to a real system of learning. Every organization calls on feedback loops that keep each component informed of the results of its own operation and consequently adjusts it in the general interest.[32]

Finally, Derrida observes, it is neither possible to dissociate "message" from "feedback regulation" nor to consider the latter as a secondary and dependent function with respect to the message that it would itself determine. "Message" and "feedback regulation," as selective functions that permit the construction and conservation of the system with respect to general entropy, imply one another:

> As I was saying earlier, we do not here have two concepts (message plus genetic regulation): in the message there was a selection or sorting and the principle of this selection constitutive of the very operation of the message had to obey economic laws. So, when Jacob changes paragraphs and objects in order to examine what he calls the second concept "as to retroaction," he is merely making the same concept of message explicit or, conversely, when he was examining the message, he was implying the retroaction. The latter consists in reintroducing in the system the results of its past action (already a memory or an archive of messages, under one guise or another) in order to observe and correct the tendency of the mechanism to disorganization.[33]

The living system cannot be a closed system; it requires some energy from the environment in order to be structured and conserved as such. However, Jacob subscribes to this point too, but only after establishing the primacy of the message as the element that merely determines the construction of the organism and thus after excluding or reducing to a secondary and dependent factor the energetic dynamic that, instead, consists in an irreducible condition:

> Ultimately, the maintenance of a living system in good repair has to be paid for: the return to the ever unstable equilibrium leads to a deficit of surrounding organization, that is, to an increase in disorder of the total system composed of the organism and its environment. The living organism, therefore, cannot be a closed

system. It cannot stop absorbing food, ejecting waste-matter, or being constantly traversed by a current of matter and energy from outside. Without a constant flow of order, the organism disintegrates. Isolated, it dies. Every living being remains in a sense permanently plugged into the general current which carries the universe towards disorder. It is a sort of local and transitory eddy which maintains organization and allows it to reproduce.[34]

Here Derrida comes to a much more radical conclusion than Jacob. Given that the energetic dynamic is the condition of the selection of the message, the opening of the living system onto the external environment must be considered as its irreducible structural condition. In other words, it is necessary to acknowledge that the living conserves itself as such to the extent that it is open onto alterity in general and thus cannot be considered as detached from and opposed to it. According to the law of supplementarity, the outside must necessarily be understood as the inside of the inside, and the other must be understood as the inside of the same:

> This self-evidence can appear trivial, I quote Jacob here only to highlight that this structural opening of any living system makes unsustainable statements about the bacterium that does not die because death comes to it from outside or about death in the proper sense that must be inscribed in the organism, etc. Just as it makes unsustainable all the simple oppositions between the inside and the outside, which underlie what the book says about sexuality and mortality as accidents come from outside to inscribe themselves in the inside. Supplementarity is inscribed in the very definition of any system, whether living or non-living.[35]

At this point, given the contradiction in which Jacob's discourse remains captured, it becomes more and more evident why we must interpret the genesis and structure of the living in terms of general textuality, as does it become more and more evident what conditions the elaboration of this genesis and structure, namely, arche-writing. Recall that Derrida elaborated this notion in order to account for the genesis of the retentional or mnestic trace and, in particular, for the problem of the selection of the retention, that is, of the "preference" in facilitation (*Bahnung*). Recall as well that, to solve these problems, he had to abandon the phenomenological perspective and go back to the natural, biologico-evolutionary conditions of the genesis of consciousness. For this reason he approached Freud and in particular Freud's early work on the genesis of the psychic system in which the psychoanalytic perspective is anchored to the biologico-evolutionary perspective of his time. Through the reading of Freud, iterability, repetition, and reproduction appeared as the structural condition of the trace—without which there would be no trace—and then the latter's genesis, motivated by the urgency of life, by the struggle for the survival, appeared as an effect of the difference of forces at work between the impression and the surface of inscription and not of the imprint left by a present and living impression. Furthermore, once elaborated in these terms, the genesis of the retentional trace has already and necessarily implied the possibility of the "feedback regulation"—that is, of the permanent reorganization of the archive of traces that constitutes the psychic system—in view of responding to contingent exigencies determined by the relationship between the psychic system and the internal and external environment:

> There is no present text in general, and there is not even a past present text, a text which is past as having been present. The text is not conceivable in an originary or modified form of presence. The unconscious text is already a weave of pure traces, differences in which meaning and force are united, a text nowhere present, consisting of archives which are *always already* transcriptions. Originary prints. Everything begins with reproduction. Always already: repositories of a meaning which was never present, whose signified presence is always reconstituted by deferral, *nachtrtäglich*, belatedly, supplementarily: for the *nachträglich* also means supplementary. The call of the supplement is primary, here, and it hollows out that which will be reconstituted by deferral as the present. The supplement, which seems to be added as a plenitude to a plenitude, is equally that which compensates for a lack. "Suppleer: 1. To add what is missing, to supply a necessary surplus," says *Littré*, respecting, like a sleepwalker, the strange logic of that word. It is within its logic that the possibility of deferred action should be conceived, as well as, no doubt, the relationship between the primary and the secondary on all levels. . . . That the present in general is not primal but, rather, reconstituted, that it is not the absolute, wholly living form which constitutes experience, that there is no purity of the living present, such is the theme, formidable for

metaphysics, which Freud, in a conceptual scheme unequal to the thing itself, would have us pursue. This pursuit is doubtless the only one which is exhausted neither within metaphysics nor within science.[36]

In "Freud and the Scene of Writing," arche-writing, as the condition of the genesis and structure of the psychic system, is already found at work as the specific differential articulation of différance that makes possible all that is and thus also the life of the living and its evolutionary history. At the time, Derrida alluded to the constant interest manifested by Freud in the articulation between psychoanalysis and the life sciences, especially apropos of *Beyond the Pleasure Principle*. It is precisely from this perspective that, according to the programme of the time, as we shall see, Derrida devotes the final part of the seminar to a systematic reading of this Freudian text, which will be further developed in the text published as "To Speculate—On Freud." The results of the seminar and, more generally, of the deconstruction of the life/death opposition, and thus of the very possibility of accounting for différance as the condition of the differential/differing relationship (between) life/death, depend on this reading.

However, before following Derrida along this line, I first draw together some conclusions about the question of the model, particularly with respect to the circular and reflexive structure of the model that is implied in the construction of the cybernetic model. As we saw, for Jacob, the cybernetic model is a good heuristic model because cybernetics elaborated the abstract notion of information as the basic component of all organized systems, in the machine as well as in the living organism and in society. It was paradoxical that, in constructing this notion to describe the logic of the living, cybernetics adopted each of these "objects," the machine and the living, as reference models to account for certain particular features of the theory of information and of organized systems. Machine, living being, and society are models for the elaboration of the model that must account for machine, living being, and society as organized systems. Derrida focuses on this paradoxical circularity or reflexivity of the model when Jacob reformulates it in greater detail, that is, apropos of the animal/machine relation that cybernetics would allow us to reconsider under a new light. In this context, there is an explicit reference to Wiener:

> With the possibility of carrying out mechanically a series of operations laid down in a programme, the old problem of the

relations between animal and machine was posed in new terms. "Both systems are precisely parallel in their analogous attempts to control entropy through feedback," said Wiener.[37] Both succeed by disorganizing the external environment, "by consuming negative entropy," to use the expression of Schrodinger and Brillouin. Both have special equipment, in fact, for collecting at a low energy-level the information coming from the outside world and for transforming it for their own purposes. In both cases, it is the realization, not the intention, that adjusts the action of the system on the outside world through the intermediary of a regulatory centre. . . . Animal and machine, each system then becomes a model for the other.[38]

As we can recall, Derrida had already remarked, in *Of Grammatology*, how strange this reflexive and circular relation between animal and machine, proposed by Wiener in *The Human Use of the Human Being*, was. If the animal and the machine is each the model of one another, then it makes no sense to speak of a model; it would be necessary instead to dismiss the traditional opposition between them and to understand their relation in a different way. However, the opposition is also the only criterion that allows us to think their relationship in terms of model, similitude, analogy. Indeed, Jacob seems to be interested in the dismissal of the opposition as much as in its conservation, which is presupposed by the construction of the cybernetic text as the model of the living. This is evident where, apropos of the reproductive activity of bacteria, he recurs again and in an explicit fashion to the analogy of the living and the machine, namely, to the factory as the model for understanding the activity of bacteria:

> If analogy is to be used, the bacterial cell is obviously best described by the model of a miniaturized chemical factory. Factory and bacterium only function by means of energy received from the exterior. Both transform the raw material taken from the medium by a series of operations into finished products. Both excrete waste products into their surroundings. But the very idea of a factory implies a purpose, a direction, a will to produce—in other words, an aim for which the structure is arranged and the activities are coordinated. What, then, could be the aim of the bacterium? What does it want to produce that justifies its existence, determines its organization and underlies

its work? There is apparently only one answer to this question. A bacterium continually strives to produce two bacteria. This seems to be its one project, its sole ambition. . . . If the bacterial cell is to be considered as a factory, it must be a factory of a special kind. The products of human technology are totally different from the machines that produce them, and therefore totally different from the factory itself. The bacterial cell, on the other hand, makes its own constituents; the ultimate product is identical with itself. The factory produces; the cell reproduces.[39]

In this case, the analogy machine/living being, factory/bacteria, is relative. Factory and bacteria are similar for their functional organization but absolutely different in the respective typologies of production: only the living being can reproduce itself and, for Jacob, as we saw, self-reproduction consists in the irreducible essence of the living, what distinguishes the living as such and opposes it to the non-living. Hence, the machine is not a good heuristic model to understand the living. However, for Derrida, it is precisely through this contradiction that the function of the model emerges, that is, the strategy to which the construction of the cybernetic text as the model of the living and, more generally, the construction and function of the model *tout court* respond:

Each system becomes a model for the other, the animal and the machine will respectively become the model of their model, which annuls—here I am also thinking of the circular ring [*anneau circulaire*]—the function of model, supposing that this function has ever existed and that this circulation does not reveal somewhere the very logic of any appeal to a model, which perhaps has always and everywhere tended to take this circular form, where the model must become the model of its model, the teleological or final sense of the model governing the mechanical or technical sense of the constructed model which becomes in turn the miniaturized or gigantic model of the finalized model, with a natural purposiveness, as this circulation is an effect of the unconceivable logic of re-production we spoke about last week, of production as starting out in reproduction.[40]

The construction of the model is teleologically oriented. In other words, it is determined by that of which the model should be the model, by that which the model would allow us to determine, and from which, precisely in view of this determination, the model must be demarcated. In this case, the cybernetic model is constructed on the basis of self-reproduction as the essence of the living, that is, it must provide a model of the organization of the living, but at the same time must be distinct, in order to legitimate the interpretation of self-reproduction as the essential trait of the living with respect to the non-living, to which the living remains necessarily opposed. Ultimately, self-reproduction as the essence of the living consists in the end that orients the selective construction of the cybernetic model and not in the result that the cybernetic model allows us to obtain. In particular, the model constructed in this way, that is, oriented to safeguard the essence of the living and its opposition to the non-living, prevents us from seeing what makes it possible, what allows us to refer to the empirical text and to other artificial products of human activity, as the analogical model of the logic of the living, and consists in a necessary consequence of the dynamic of self-reproduction, but only from the perspective of general textuality and arche-writing. The empirical text, the machine, or the factory can be taken up as models of the living as they are products of the living, specific differential articulations of the self-reproduction that structures the living according to the dynamic of general textuality and arche-writing. They are effects, even if they are distant and specific, of the self-reproduction that structures the living and that, thanks to its irreducible structural opening, cannot but be propagated, transmitted to its products, with the unavoidable consequence that differential/differing self-reproduction cannot be considered as the exclusive essence of the living and only of the living:

> There is no model for reproduction except the model of the model or reproduction, the model of model or reproduction itself. If the genetic message "resembles a text without an Author [*comme un texte sans auteur*], that a proof-reader has been correcting for more than two billion years, continually improving, refining and completing it, gradually eliminating all imperfections,"[41] here the concept of text is not a model or an analogy: firstly, because what we understand as a "text without author" (in the current sense, imagined as a book or a manuscript without signature in nature or in a library) is already a product [*produit*], and thus a re-product [*re-produit*], an effect of the living as genetic message and, therefore, its structure is already inseparable from the structure of the living; on the other hand, because if the text,

which today best resembles the genetic message, that is, the structure of the living, cannot simply have the status of model, this is because there has never been a model for the living. It is this as it were internal deconstruction (internal and supplementary) of the concept of model that intervenes when Jacob has recourse to the concept of text and when he recognizes that "the genetic message can be translated only by the very products of its translation," that is, that, ultimately, no translation is possible: textuality cannot be absolutely translated, despite all the effects of translation it induces.[42]

V

Between Life and Death: The Bond

The last four sessions of the seminar *La vie la mort* are devoted to the reading of Freud's *Beyond the Pleasure Principle* (1920).[1] It is well known that the father of psychoanalysis develops in this text a critical examination of the pleasure principle, which he had posited as the mover of drive activity and thus as the constitutive factor of psychic life. In the first part of the text, Freud takes into account a series of psychic phenomena that seem to contradict the pleasure principle. In particular, he finds in the tendency to repeat painful or traumatic events, which is examined in the singular case of the ludic activity of a child (his grandson Ernst playing with the "wooden spool"), in the neuroses of war, in dreams and transference phenomena in some of his patients in analysis, a tendency that is independent from the pleasure principle, which is defined as "repetition compulsion." Therefore, he is forced to interrupt the examination of the concrete data of analytic observation and to venture onto the ground of merely theoretical speculation, which will lead him to introduce the death drives as the manifestation of the most general law that regulates the life of the living. All forms of life would tend to the restoration of the state of balance that characterizes inert matter and thus to death. Within this speculative horizon, Freud appeals to the most various textual resources: not only to psychoanalysis but also to literature (Goethe's *Faust*, Schiller, Rückert), more or less explicit autobiographic narrative, private correspondence, philosophy (Nietzsche and Schopenhauer, through the denegation of their legacy, specifically Fechner and Plato), Greek and Eastern mythology, the laws of thermodynamics and chemistry, embryology, neurology, physiology, evolutionary theory, and, above all, the biology of reproduction and of hereditariness (Hartmann, Goette, Hering, and, in particular, Weismann), leading the text to a "metaphorical" drift that cannot be arrested and in which the recourse to a politico-military terminology is privileged to describe psychic life as well as the struggle for hegemony, mastery, and sovereignty (*Herrschaft*). Therefore, from the perspective of its composition, *Beyond the Pleasure Principle* is an example or an effect of this general textuality that, weaving the life of the living as a tissue of iterable traces, of differential references, necessarily affects its activity, too, and thus the texts that the latter produces. This may be verified in an exemplary way, as Derrida remarks, when those texts treat the very object of which they are a result—that is, life itself. It is for this reason that Derrida devotes special attention to those elements that attest to the implication of Freud's life in the composition of *Beyond the Pleasure Principle*: the play of his grandson Ernst as the example of repetition compulsion, and the death of Sophie, the beloved daughter of the father of psychoanalysis and Ernst's mother. It is Freud himself who takes into account the possibility that the death of Sophie, which occurs during the drafting of the text, influenced his theses in a consolatory and conciliatory direction. Freud will explicitly deny this possibility, but nothing excludes that it is a denegation, especially from the perspective of general textuality. This aspect is present in the seminar *La vie la mort* but will be developed only in the version published in *The Post Card*. For this reason, and consistent with the coordinates of this work, I do not directly address this feature but instead devote my analysis to the speculation on the living and the weave of speculation and biology. As we shall see, Derrida understands Freud's biological speculation as thematically congruent with the logic of the living developed by Jacob, particularly as regards the conception of sexuality and death as factors that intervened only recently during the evolution of life and, above all, as regards the tendency of the living to die a proper death, internal, immanent to the living, and thus to elude death due to factors considered to be external, environmental, and contingent. But before approaching this articulation of Jacob with Freud, which is decisive from the perspective of this work, I trace and reconstruct first the path described by Derrida through the reading of *Beyond the Pleasure Principle* and focus on the development of the apparently enigmatic passage from "Freud and the Scene of Writing" I quoted in the first chapter and which I interpreted as a key trace in view of the deconstruction of the metaphysical conception of life and thus of the latter's reelaboration into différance and arche-writing:

All these differences in the production of the trace may be reinterpreted as moments of deferring [*différance*]. In accordance with a motif which will continue to dominate Freud's thinking, this movement is described as the effort of life to protect itself by *deferring* a dangerous cathexis, that is, by constituting a reserve (Vorrat). The threatening expenditure or presence is deferred with the help of breaching or repetition. Is this not already the detour (Aufschub, lit. delay) which institutes the relation of pleasure to reality (*Beyond the Pleasure Principle*)? Is it not already death at the origin of a life which can defend itself against death only through an economy of death, through deferment, repetition, reserve?[2]

No doubt life protects itself by repetition, trace, différance (deferral). But we must be wary of this formulation: there is no life present *at first* which would then come to protect, postpone, or reserve itself in différance. The latter constitutes the essence of life. Or rather: as différance is not an essence, as it is not anything, *it is not* life, if Being is determined as *ousia*, presence, essence/existence, substance, or subject. Life must be thought of as trace before Being may be determined as presence. This is the only condition on which we can say that life *is* death, that repetition and the beyond of the pleasure principle are native and congenital to that which they transgress.[3]

It is worth noting that, in the first session of the seminar devoted to Freud, Derrida refers explicitly to "Freud and the Scene of Writing," as well as to *Le Facteur de la verité* and to those parts of *Glas* dedicated to fetishism, as the essential premises of the reading of *Beyond the Pleasure Principle* he is going to mobilize.[4]

Beyond the Pleasure Principle: Différance

Freud begins with summarizing the assumptions of psychoanalysis with regard to the hegemony of the pleasure principle in psychic life. He introduces an economic point of view that allows him to account for the energetic and not merely the structural dimension of the pleasure principle:

> In the Theory of psycho-analysis we have no hesitation in assuming that the course taken by mental events is automatically regulated by the pleasure principle. We believe, that is to say, that the course of those events is invariably set in motion by an unpleasurable tension, and that it takes a direction such that its final outcome coincides with a lowering of that tension, that is, with an avoidance of unpleasure or a production of pleasure. In taking that course into account in our consideration of the mental processes which are the subject of our study, we are introducing an "economic" point of view into our work; and if, in describing those processes, we try to estimate this "economic" factor in addition to the "topographical" and "dynamic" ones, we shall, I think, be giving the most complete description of them of which we can at present conceive, and one which deserves to be distinguished by the term "metapsychological."[5]

Given the insufficiency of the definitions of pleasure inherited from the philosophical and psychological traditions, Freud proposes a definition that essentially takes up the economic terms and the neuro-physiological arguments he had elaborated at the time of the *Project*: "We have decided to relate pleasure and unpleasure to the quantity of excitation that is present in the mind but is not in any way 'bound'; and to relate them in such a manner that unpleasure corresponds to an *increase* in the quantity of excitation and pleasure to a *diminution*."[6] At this point, we encounter a preliminary, obvious objection: the affirmation of the hegemony of the pleasure principle clashes with incontrovertible evidence: the displeasure that characterizes several phenomena of psychical life. However, psychoanalysis had already dealt with this objection and had responded by introducing the reality principle:

> The most that can be said, therefore, is that there exists in the mind a strong *tendency* towards the pleasure principle, but that that tendency is opposed by certain other forces or circumstances, so that the final outcome cannot always be in harmony with the tendency towards pleasure. . . . We know that the pleasure principle is proper to a *primary* method of working on the part of the mental apparatus, but that, from the point of view of the self-preservation of the organism among the difficulties of the external world, it is from the very outset inefficient and even

highly dangerous. Under the influence of the ego's instincts of self-preservation [*Selbstheraltungstriebe*], the pleasure principle is replaced by the *reality principle*.[7]

The search for a direct and immediate satisfaction of the pleasure principle, which the primary drive processes express, may expose the individual to threats coming from the external environment and thus put life in danger. The "reality principle," the expression of the individual's drives of self-conservation, allows the individual to escape these threats by deferring the satisfaction of pleasure in time; displeasure would be the feeling that accompanies this deferral of pleasure until satisfaction:

> This latter principle does not abandon the intention of ultimately obtaining pleasure, but it nevertheless demands and carries into effect the postponement [*Aufschub*] of satisfaction, the abandonment of a number of possibilities of gaining satisfaction and the temporary toleration of unpleasure as a step on the long indirect road to pleasure [*auf dem langen Umwege zur Lust*]. The pleasure principle long persists, however, as the method of working employed by the sexual instincts [*Sexualtriebe*], which are so hard to "educate," and, starting from those instincts [*Triebe*], or in the ego itself, it often succeeds in overcoming the reality principle, to the detriment of the organism as a whole.[8]

Derrida explicitly interprets this deferral as an effect of différance, which is understood, as we saw in "Freud and the Scene of Writing," as the deferral that presupposes the act of putting into reserve as its condition of possibility and, at the same time, produces it. Hence, the very principle of pleasure is an effect of différance as it institutes and conserves itself by differing (from) itself, through the reality principle:

> The reality principle imposes no definitive inhibition, no renunciation of pleasure, only a detour in order to defer enjoyment, the waystation [*le relais*] of a *différance (Aufschub)*. On this "long indirect road" (*auf dem langen Umwege zur Lust*) the pleasure principle submits itself, provisionally and to a certain extent, to its own lieutenant. The latter, as representative, slave, or informed disciple, the disciplined one who disciplines also plays the role of the preceptor in the master's service.[9]

Therefore, the reality principle cannot be interpreted as the other and the opposite of the pleasure principle but must be understood as the effect of the self-differing that affects and structures the pleasure principle and thus psychic life, and not only that the reality principle grants the survival of the organism in the environment.

> As soon as an authoritarian agency submits itself to the work of a secondary or dependent agency (master/slave, master/disciple) which finds itself in contact with "reality"—the latter being defined by means of the very possibility of this speculative transaction—there is no longer any *opposition, as* is sometimes believed, between the pleasure principle and the reality principle. It is the same *differant, in différance* with itself. But the structure of *différance* then can open onto an alterity that is even more irreducible than the alterity attributed to opposition. Because the pleasure principle—right from this preliminary moment when Freud grants it an uncontested mastery—enters into a contract only with itself, reckons and speculates only with itself or with its own metastasis, because it sends itself *(il s'envoie)* everything it wants, and in sum encounters no opposition. It *unleashes* in itself the *absolute* other.[10]

We start figuring, beyond Freud, the differential and non-oppositional dynamic that, according to the general coordinates of the seminar, allows us to deconstruct the traditional life/death opposition and, at the same time, to account for their irreducible co-implication:

> Pure pleasure and pure reality are ideal limits, which is as much as to say fictions. The one is as destructive and mortal as the other. Between the two the *differant* detour therefore forms the very actuality [*effectivité*] of the process, of the "psychic" process as a "living" process. Such an "actuality," then, is never present or given. . . . Therefore one cannot even speak of effective actuality, of *Wirklichkeit*, if at least, and in the extent to which, it is coordinated with the value of presence. The detour thereby "would be" the common, which is as much as to say the *differant* root of the two principles, the root uprooted from itself, necessarily impure, and structurally given over to compromise, to the speculative transaction. The three terms—two principles

plus or minus *différance*—are but one, the same divided, since the second (reality) principle and *différance* are only the "effects" of the modifiable pleasure principle. But from whichever *end* one takes this structure with one-two-three terms, it is death. *At the end*, and this death is not opposable, does not differ, in the sense of opposition, from the two principles and their *différance*. It is inscribed, although non-inscribable, in the process of this structure—which we will call later stricture. If death is not opposable it is, already, *life death*.[11]

The psychic process as a living process is structured to escape death, to defer it for as long as possible; therefore, death is an irreducible condition of the structural possibility of the living process. For this reason, it cannot be considered as the other opposed to life but rather as the condition of the self-differing that structures the living process and thus as internal to life. The living lives and thus escapes death as much as it differs (from) itself; but if it must differ from itself in order to differ itself (that is, to survive and conserve itself for a certain interval of time), if it must be other than itself, alter itself, then alterity in general, as well as death, consist in the irreducible condition of this twofold differing and thus of life itself. Hence, we can no longer speak of life *and* death as they can no longer be considered external to one another, determined by themselves and thus opposable, but we must recur to the neologism *life/death* to describe the differential/differing dynamic as the structure of the living process, at least until we invent a term that accounts for this process in a more economical way. It is not by chance that Derrida refers here to Blanchot's *Death Sentence* (*L'arrêt de mort*), to the double meaning—death sentence/suspension of death—and thus to "Living On," which was in press at the time of the seminar but appeared just before *The Post Card*.[12] Nor is it by chance that he raises the question of survival, which I address at the end of this volume. From this perspective, Derrida explicitly affirms that his interpretation goes far beyond Freud, or at least beyond the letter of Freud's text:

> This Freud does not say, does not say it presently, here, nor even elsewhere in this form. It gives (itself to be) thought without ever being given or thought. Neither here nor elsewhere. But the "hypothesis" with which I read this text and several others would go in the direction of disengaging that which is engaged here between the first principle and that which appears as *its*

other, to wit, the reality principle as *its* other, the death drive as its *other:* a structure of alteration without opposition. That which seems, then, to make the belonging—a belonging without interiority—of death to pleasure more continuous, more immanent, and more natural too, also makes it more scandalous as concerns a dialectics or a logic of opposition, of position, or of thesis. There is no thesis of this *différance.* The thesis would be the death sentence *(arrêt de mort)* of *différance.* The syntax of this *arrêt de mort,* which arrests death in two *différants* senses (*a* sentence which condemns to death and an interruption suspending death) will be in question elsewhere [in *Survivre,* à paraître].[13]

In particular, distancing himself from Freud, for whom only the direct and exclusive affirmation of the pleasure principle implies death, Derrida contends that death would also be the consequence of the direct and exclusive affirmation of the reality principle as well as of pure différance. This means that the "life" of the living is instituted, conserved, and lasts until the differential relation—the transaction—between the two principles is conserved and lasts, a relation for which life is a relative but still determined effect. Pure différance, differing as such, loosened and independent from the two principles that structure the living process, necessarily brings about death as it gives no place to any stable determination of the process of which it is in any case an irreducible but immanent condition:

Each time that one of the "terms," the pseudo-terms or pseudopods, sets forth [*marche*] and goes *to the end* [*au-bout*] of itself, and therefore of its other, keeping to its extreme and pure autarky, without negotiating, without speculating, without passing through the mediation of any third party [*du tiers*], it is death, the mortal sprain which puts an end to the strain of calculation. If the reality principle autonomizes itself and functions all alone (an absurd hypothesis by definition, covering the field said to be pathological), it cuts itself off from all pleasure and all desire, from the entire auto-affective relation without which there is neither any desire nor pleasure that can appear at all. This is the sentence of death [*arrêt de mort*], of a death that is also at the two other ends: equally in the fact that the reality principle then would affirm itself without any erotic enjoyment, and in the other fact that it would be the death of its service, its delegated service of the pleasure principle. It would die itself, in its ordered service, due to the economic zeal of pleasure, of a pleasure too jealous of itself and of what it sets aside. It would already be pleasure that, by itself protecting itself too much, would come to asphyxiate itself in the economy of its own reserves. But inversely (if it can be put thus, for this second eventuality does not invert the first one), to go to the end of the transactional compromise that is the *Umweg*—*pure différance* in a way—is also the *arrêt de mort:* no pleasure would ever present itself. But does a pleasure ever present itself? Death is inscribed, although noninscribable, "in" *différance* as much as it is in the reality principle which is but another name for it, the name of another "moment," since pleasure and reality are also exchanged within it.[14]

Therefore, concluding his reading of the first chapter of *Beyond the Pleasure Principle*, Derrida finds in the function ascribed to the pleasure principle—that of keeping cathexis as low as possible, or at least constant—the death drives that Freud brings to the stage only in the second part of *Beyond the Pleasure Principle* and thus at the moment in which his observations about repetition compulsion will lead him to advance the hypothesis of the existence of independent drives, more originary than the pleasure principle, and thus to venture into the field of biological speculation on the origin and evolution of life:

It is only in chapter IV, announcing the speculation of great breadth, that Freud envisages a function of the psychic apparatus which, without being *opposed* to the PP [Pleasure Principle] would be no less independent from it, and more originary than the tendency (as distinct from *the* function) to seek pleasure and to avoid unpleasure: the first exception before which, in sum, "speculation" would never have begun. . . . Thus, the speculative overflowing still awaits. And the great breadth. It will *lead* to another "hypothesis": drives "in the service of which" the absolute master, the PP, would work. The drives said to *be* of death. But were they not *already* at work in the logic we have just recognized?[15]

However, as we shall see, Freud himself admits toward the end of *Beyond the Pleasure Principle* that the function attributed to the pleasure principle is one of the motives, if not the only one, that led him to hypothesize the existence of the death drives. At this point, for Derrida, it is necessary to follow the speculative transitions through which Freud arrives at this conclusion, as if it were the *telos* that secretly and silently orients the development, and thus to detect beyond Freud the structural and structuring effects that the differential/differing dynamic produces as the irreducible condition of the life (death) of the living.

Différance and Stricture: The Bond

In Chapter IV, Freud explores the paths of speculation to account for the tendency to repetition that characterizes the oneiric life in the cases of traumatic neurosis. These cases, in which the dream repeats the traumatic event, contradict the most general analytic assumption according to which the dream is the satisfaction of an unfulfilled desire, and thus they seem to testify to a psychic activity that is independent from the pleasure principle. Freud begins with recalling the description of the system *Pcpt.-Cs.* (perception-consciousness), a formula elaborated in the writings on *Metapsychology* (1915), but already suggested in the *Interpretation of Dreams* (1899). To describe the specificity of that system with respect to the other systems that constitute the psychic apparatus—the unconscious as well as the conscious—Freud conjures up the notion of facilitation (*Bahnung*), elaborated at the time of the *Project*. External stimuli do not leave permanent traces in the *Pcpt.-Cs.* but are conserved in other systems as mnemic residua that constitute the basis of memory. Venturing into speculation but supported by the scientific knowledge of the time, Freud proposes a speculative fiction to explain the organic genesis of the system *Pcpt.-Cs.* from a biologico-evolutionary perspective:

> Let us picture a living organism in its most simplified possible form as an undifferentiated vesicle of a substance that is susceptible to stimulation. Then the surface turned towards the external world will from its very situation be differentiated and will serve as an organ for receiving stimuli. Indeed embryology, in its capacity as a recapitulation of developmental history, actually shows us that the central nervous system originates from the ectoderm; the grey matter of the cortex remains a derivative of the primitive superficial layer of the organism and may have inherited some of its essential proprieties.[16]

The ceaseless exposure to stimuli would produce a cortex on the external surface of the vesicle that makes the reception of stimuli possible and prevents further modifications. The stimuli coming from without would encounter some resistance only when penetrating into the external cortex; once reaching a certain depth, they would leave a trace where the resistance is overcome:

> It may be supposed that, in passing from one element to another, an excitation has to overcome a resistance, and that the diminution of resistance thus effected is what lays down a permanent trace of the excitation, that is, a facilitation [Bahnung]. In the system Cs., then, resistance of this kind to passage from one element to another would no longer exist. This picture can be brought into relation with Breuer's distinction between quiescent (or bound) and mobile cathectic energy in the elements of the psychical systems; the elements of the system Cs. would carry no bound energy but only energy capable of free discharge.[17]

It is worth paying attention to the first reference to Breuer's model, whose source, particularly for what concerns the distinction between free energy and bound or tonic energy, is Hermann von Helmoltz's work on thermodynamics titled *Über die Thermodynamik chemischer Vorgänge* (1882).[18] Derrida recuperates Helmoltz's definition, quoted in Laplanche's *Life and Death in Psychoanalysis* [*Vie et mort en psychanalyse*, 1970], the reading of which was required in Derrida's seminar:

> It seems certain to me that we must distinguish, within chemical processes as well, between that portion of the forces of affinity capable of being freely transformed into other kinds of work, and that portion that can only become manifest in the form of heat. To abbreviate, I shall call these two portions of energy: free energy and bound energy.[19]

So far there is nothing new, at least with respect to the *Project*. However, Freud goes on with his biologico-speculative hypothesis in order to explain that the external surface of the vesicle is not only receptive with respect to stimuli but is also a "protective shield";[20] otherwise, it could not bear the quantity and intensity of the stimuli coming from the external world and thus would succumb. For this reason, in higher organisms the cortex would evolve, developing into the receptive sensorial apparatus that filters stimuli into the brain that receives them by withdrawing to a more internal and protected position. However, this sensorial apparatus, "which is later to become the system *Pcpt-Cs*,"[21] receives stimuli not only from the outside but also from within, and with regard to the latter it has no protective shield. The stimuli from within "extend into the system directly and in undiminished amount."[22] Therefore, in the case of traumatic stimuli coming from the outside, able to break the protective shield, the psychic system is invested with a great amount of potentially harmful excitation and thus confronts the urgency "of mastering the amounts of stimulus which have broken in and of binding them, in the psychical sense, so that they can be disposed of."[23] Once this protective shield is broken, the excitation caused by the trauma pours freely and in large amounts into the psychic apparatus as if it came from within the apparatus itself: the latter reacts by investing portions of its energy to "bind" the free-flowing excitation:

> And how shall we expect the mind to react to this invasion? Cathectic energy is summoned from all sides to provide sufficiently high cathexes of energy in the environs of the breach. An "anticathexis" on a grand scale is set up, for whose benefit all the other psychical systems are impoverished, so that the remaining psychical functions are extensively paralysed or reduced. . . . From the present case, then, we infer that a system which is itself highly cathected is capable of taking up an additional stream of fresh inflowing energy and of converting it into quiescent cathexis, that is of binding it psychically. The higher the system's own quiescent cathexis, the greater seems to be its binding force; conversely, therefore, the lower its cathexis, the less capacity will it have for taking up inflowing energy and the more violent must be the consequences of such a breach in the protectivre shield against stimuli.[24]

Freud believes he can explain the repetition compulsion that characterizes traumatic neuroses. This kind of neurosis would be caused by a sudden trauma for which the psychic system is totally unprepared. Being unprepared is the cause of the fright that characterizes the traumatic event, and it is precisely this fright, rather than the physical violence that was actually suffered, that would be the cause of neurosis. Indeed, for Freud, we must distinguish fright from anxiety: as a fear of an expected even if undetermined danger, the latter already constitutes a form of preparation to danger and thus of defense against the effects of the trauma.[25] The fact that the psychic system is unprepared for an unexpected danger implies a weak energetic investment and thus an inability (or a diminished ability) to bind the flow of excitation produced by the trauma. The repetition of the traumatic event that characterizes the dreams of the patients affected by this type of neurosis can therefore be explained as the attempt to bind, in the dream, the flow of excitation against which the psychic system was found unprepared while awake. Therefore, the hypothesis of a function of the psychic apparatus independent from the pleasure principle, so far considered as the decisive mover of psychic life, would be confirmed:

> The fulfillment of wishes is, as we know, brought about in a hallucinatory manner by dreams, and under the dominance of the pleasure principle this has become their function. But it is not in the service of that principle that the dreams of patients suffering from traumatic neuroses lead them back with such regularity to the situation in which the trauma occurred. We may assume, rather, that dreams are here helping to carry out another task, which must be accomplished before the dominance of the pleasure principle can even begin. These dreams are endeavouring to master the stimulus retrospectively, by developing the anxiety whose omission was the cause of the traumatic neurosis. They thus afford us a view of a function of the mental apparatus which, though it does not contradict the pleasure principle, is nevertheless independent of it and seems to be more primitive than the purpose of gaining pleasure and avoiding unpleasure.[26]

The repetition compulsion reveals a function of the psychic apparatus—mastering dangerous stimuli and excitation—that is independent from the pleasure principle but not in contradiction with the latter, as it precedes and makes the institution of the pleasure principle possible. This function consists in the genetico-structural condition of possibility of the pleasure principle. Derrida emphasizes the politico-military metaphorical register that supports and orients the Freudian description and, consistently with Freud's argument, isolates the condition of this function: it essentially consists in binding the excitation that could harm the psychic by freely pouring in it, that could put it out of use or, more simply, like in the case of pure différance, prevent it from being constituted as such:

> This is the first exception to the law according to which the dream fulfills a wish. But this law is not "contradicted," the

exception does not speak *against* the law: it precedes the law. There is something older than the law within the law. The law could appear to govern the function of the dream only after the institution of the PP in its dominance. This latter therefore would be a relatively late effect of a history, of an original genesis, a prior victory on a field that does not belong to the PP in advance, and of which the PP is not even a native [*autochtone*]: victory and capture, binding triumphs over unbinding, the band over the contra-band, or even the contra-band over the a-band or the disband. Over absolute astricture, if some such thing could take place and shape.²⁷

It is Freud who draws this conclusion when he describes the role that the function of mastery carries out in psychic life in general and thus with respect to the drives of the organism that affect the psychic system from within, that is, without encountering a protective shield or other defensive systems. These drives represent the effect of the unconscious "primary processes" that consist of "at once the most important and the most obscure element of psychological research,"²⁸ because it is with respect to these drives that the psychic life of the individual is structured:

I described the type of process found in the unconscious as the "primary" psychical process, in contradistinction to the "secondary" process which is the one obtaining in our normal waking life. Since all instinctual impulses [*Triebregungen*] have the unconscious system as their point of impact, it is hardly an innovation to say that they obey the primary process. Again, it is easy to identify the primary psychical process with Breuer's freely mobile cathexis and the secondary process with changes in his bound or tonic cathexis. If so, it would be the task of the higher strata of the mental apparatus to bind the instinctual excitation [*Erregung der Triebe*] reaching the primary process. A failure to effect this binding would provoke a disturbance analogous to a traumatic neurosis; and only after the binding has been accomplished would it be possible for the dominance of the pleasure principle (and of its modification, the reality principle) to proceed unhindered. Till then the other task of the mental apparatus, the task of mastering or binding excitations, would have precedence—not, indeed, in *opposition* to the pleasure principle, but independently of it and to some extent in disregard of it.[29]

However, for Derrida, Freud does not seem to realize the possible consequences of his speculation, which is to say, he does not seem to develop the resources implicit in it. Freud's text allows us to find in the exercise of the binding function of the primary processes a function independent from the pleasure principle, beyond the interpretation given by Freud himself, the effect of a more originary psychic activity than the one testified to by the pleasure principle:

> This obscurity, which Freud does not insist upon, is due to the fact that before the instituted mastery of the PP there is *already* a tendency to binding, a mastering or stricturing impulse that foreshadows the PP without being confused with it. It collaborates with the PP without being of it. A median, *differing* [différante] *or indifferent* zone (and it is differing only by being indifferent to the oppositional or distinctive difference of the two borders), relates the primary process in its "purity" to the "pure" secondary process entirely subject to the PP. A *zone*, in other words a *belt* between the pp [primary process] and the PR, neither tightened nor loosened *absolutely*, everything *en différance de stricture*.[30]

As this function is independent from the pleasure principle and consists in binding the primary processes through the secondary processes that characterize ordinary life, that is to say, the state of wakefulness in which the reality principle is active, then it is first of all legitimate to suppose that a tendency to binding, to the mastery of drives, which is more originary and independent of the pleasure principle, is the condition of psychic life. Furthermore, because the pleasure principle is instituted only when the primary processes are "conveniently" bound and thus when the struggle between the forces of investment and counter-investment, between the free energy of the primary processes and the binding energy of the secondary processes, is resolved into a stable formation that the pleasure principle can master, then we can also legitimately suppose that psychic life is first conditioned by this tendency to binding, to the mastery of drives and, thus, that the latter consists in the effect of the differential/differing relation between primary processes and secondary processes, according to which the one is only in relation to the other, to the other that is different/differed. Hence, the opposition of primary and secondary processes, which presupposes their respective independence and autonomy, must be considered a theoretical fiction (a "myth") that is inaccessible, above all to the pleasure principle, which can master the psychic life only through a differential/differing relation with the reality principle. As I shall point out, Derrida will develop this resource borrowed from Freud's speculation and left unemployed—the individuation of a tendency to binding that would be more originary and independent of the pleasure principle—and will find there the "proper" of the living.

Beyond the Death Drive the Drive of the Proper

The function carried out by the repetition compulsion with respect to the primary processes leads Freud to continue with his speculation up to the hypothesis of the universal properties of drives and more generally of life itself and its evolution:

> At this point we cannot escape a suspicion that we may have come upon the track of a universal attribute of instincts [*Triebe*] and perhaps of organic life in general which has not hitherto been clearly recognized or at least not explicitly stressed. *It seems, then, that an instinct [Trieb] is an urge inherent in organic life to restore an earlier state of things* which the living entity has been obliged to abandon under the pressure of external disturbing forces; that is, it is a kind of organic elasticity, or, to put it another way, the expression of the inertia inherent in organic life.[31]

Drives tend to restore matter to the state of inertia from which life would emerge thanks to perturbing external forces; the course of life would be merely a deviation, a more or less long and complicated deviation, from that state of inertia that the drives have to restore by pushing life toward its immanent conclusion and thus toward death:

> Moreover it is possible to specify this final goal of all organic striving. It would be in contradiction to the conservative nature of the instincts [*Triebe*] if the goal of life were a state of things which had never yet been attained. On the contrary, it must be an *old* state of things, an initial state from which the living entity

has at one time or other departed and to which it is striving to return by the circuitous paths along which its development leads. If we are to take it as a truth that knows no exception that everything living dies for *internal* reasons—becomes inorganic once again—then we shall be compelled to say that *"the aim of all life is death"* and, looking backwards, that *"inanimate things existed before living ones."* . . . For a long time, perhaps, living substance was thus constantly created afresh and easily dying, till decisive external influences altered in such a way as to oblige the still surviving substance to diverge ever more widely from its original course of life and to make ever more complicated *détours* before reaching its aim of death. These circuitous paths to death, faithfully kept to by the conservative instincts [*konservativen Trieben*], would thus present us today with the picture of the phenomena of life.[32]

Derrida, in turn, speculates on this speculative hypothesis about the origins and evolution of life by investing once again in the textual resources that Freud himself left unemployed, that is, on the notion of deviation (*détour*, *Umweg*). Here he finds a confirmation of the law of différance as the condition of the differential/differing relationship of life/death, of their irreducible co-implication. If, as already seen, the differential deviation characterizes the relation between the pleasure principle and the reality principle, so that the former must differentiate/differ itself through the latter in order to be itself, then we can recognize the same dynamic in the deviation from the state of inertia that inaugurates life: life would be anything but the self-differing of death. Or, rather, we must recognize that the one is the condition of the other: if death differs (from) itself in life in order to be itself, if life is anything but this differing deviation of/from death, then this deviation that affects life in general is the condition of possibility of the particular deviation that informs the relationship between the pleasure principle and the reality principle and thus psychic life in general:

> The detour is expanding immeasurably. I mean the *Umweg*. We had already encountered, starting with the first chapter, this notion of the *Umweg*. At that point, in question were the relations between PP and PR. Here, the determination of the detour in the procedure would be more general. This determination overflows the one in the first chapter, and provides its

basis [*son assise*]. The *Umweg* would differ/defer not with the aim of pleasure or of conservation (the relay of the PR in the service of the PP), but with the aim of death, or of the return to the inorganic state. The *Umweg* of the first chapter would constitute only an internal, secondary, and conditional modification of the absolute and unconditional *Umweg*. It would be in the service of the general *Umweg*, of the (no) step of the detour [*pas de détour*] which always leads back to death. *Leads back*—here again it is not a question of going, but of coming back [*revenir*]. It is this double determination that I had assigned to the "word" *différance* with an *a*. It follows equally, then, that the *Umweg* is not a derivative type of path or step. It is not a passing determination, a narrower or stricter definition of the passage, it is the passage. (The) *Weg* (is) *Umweg* from the first step of the step.[33]

Therefore, Freud seems to allow that death is not a simple accident of life, external and opposed to it, but an internal condition. However, affirming that death as the origin and internal end of life implies that death is the essence of life and thus that life is the other through which death returns to itself, accomplishes itself according to its own end:

The end of the living, its aim and term, is the return to the inorganic state. The evolution of life is but a detour of the inorganic aiming for itself, a race to the death. It exhausts the couriers, from post to post, as well as the witnesses and the relays. This death is inscribed as an internal law, and not as an accident of life (what we had called the law of supplementarity in the margins of *The Logic of the Living*[34]). It is life that resembles an accident of death or an excess of death, in the extent to which it "dies for internal reasons" (*aus inneren Gründen*).[35]

Freud seems to take a position that is opposed to a mirror image of the position taken by Jacob, for whom death is a supplement of life and intervenes only recently in the course of evolution, and to that of Hegel, for whom, I recall, death is anything but the moment of mediation (the being other) through which life reappropriates itself in its ideal determination. A little later, Derrida explicitly affirms that "we are reading Freud with one hand, and with the other, via an analogous vocabulary, the Hegel of the dialectic of the master and the slave."[36] Indeed, for Freud the conservative drives that lead the living to extend its life do not contradict the death drive but are at the service of the latter, as they carry out the function of preventing the living from dying an accidental death, due to environmental factors and considered external to the living, and thus of granting that it dies of internal causes, immanent to life:

> Seen in this light, the theoretical importance of the instincts of self-preservation [*Selbsterhaltungstriebe*], of self-assertion [*Machttriebe*] and of mastery [*Geltungtriebe*] greatly diminishes. They are component instincts [*Partialtriebe*] whose function it is to assure that the organism shall follow its own path to death, and to ward off any possible ways of returning to inorganic existence other than those which are immanent in the organism itself. We have no longer to reckon with the organism's puzzling determination (so hard to fit into any context) to maintain its own existence in the face of every obstacle. What we are left with is the fact that the organism wishes to die only in its own fashion. Thus these guardians of life, too, were originally the myrmidons of death.[37]

Like Jacob, Freud dissociates and opposes an external, contingent death, conditioned from environmental factors, and an internal death, immanent to the living, a proper death and an improper one. The conservative drives, therefore, must assure that the deviation of which life consists ends according to the *telos* that has oriented it since the beginning and from within, that is, the return to the inert and undifferentiated state that characterizes lifeless matter. Therefore, the drives of self-preservation must assure that the living dies of its own death, that it appropriates itself by appropriating the death that belongs to it, that constitutes the internal essence of the living. Like in Hegel's dialectics, but in an overturned fashion, the drives of self-preservation consist in the moment of transition (the being other) through which death returns to itself and reappropriates itself. Hence, Derrida draws a consequence that leads well beyond Freud and allows him to uncover the deeper motive, the hidden desire, which would have induced Freud to the formulation of the hypothesis of the death drive: the more general law to which drives are subordinated is the drive of the proper or the appropriation drive:

The drive of the proper would be stronger than life *and* than death. We must, then, unfold the implications of such a statement. If, auto-teleguiding its (his) own legacy, the drive of the proper is stronger than life and stronger than death, it is because, neither living nor dead, its force does not qualify it otherwise than by its own, proper drivenness, and this drivenness would be the strange relation to oneself that is called the relation to the proper: the most driven drive is the drive of the proper, in other words the one that tends to reappropriate itself. The movement of reappropriation is the most driven drive. The proper of drivenness is the movement or the force of reappropriation. The proper is the tendency to appropriate oneself.[38]

However, if the living must go through the deviation of which life consists, in order to appropriate itself through its own death, then it has always already been expropriated from itself and is not itself unless it differs (from) itself. It is precisely because the living must appropriate itself in order to be itself that the differential deviation (*Umweg* or différance) must be considered the irreducible condition of possibility of self-appropriation and, at the same time, what makes the self-return impossible:

All the *différance* is lodged in the desire (desire is nothing but this) for this auto-tely. It auto-delegates itself and arrives only by itself differing/deferring itself in (its) totally-other. No more proper name, no proper name that does not call (to) itself, or call upon this law of the *oikos*. In the guarding of the proper, beyond the opposition life/death, its privilege is also its vulnerability, one can even say its essential impropriety, the exappropriation (*Enteignis*) *which* constitutes it.[39]

On the other hand, Freud already seems to admit that the accomplishment of this drive process and thus of self-appropriation is impossible, at least for the human being, so long as, in that case, the primary drives are repressed and thus their satisfaction is indefinitely differed.[40] Therefore, Derrida concludes:

Therefore the *exappropriating* structure is irreducible and undecomposable. It redirects repression [*refoulement*]. It always prevents reappropriation from closing on itself or from achieving itself in a circle, the economic circle or the family circle. . . . The repressed drive *"ungebändigt immer vorwärts dringt"*: undisciplined, refractory, untamed, never permitting itself to be bound or banded by any master, it always pushes forward. It is that the backward path *(Der Weg nach rüickwärts . . .)* is always both displaced and "obstructed" *(verlegt)* by a repression. The latter does not affect the *Weg* or the *step* of the outside, it is its very proceeding and in advance finds itself *unterwegs*, en route.[41]

Furthermore, Freud allows that the hypothesis of the death drive and the very idea that death is proper to life, internal and immanent to it, may be the result of a hidden desire, more or less conscious, the desire for consolation before what, in fact, remains unappropriable: death, and thus life. Before this risk, Freud turns to the support of genetic biology, searching for a confirmation of his speculative theory and in particular of the hypothesis about the death drive.[42] For Derrida, here we can find a point of extreme congruency, which is not only thematic, between Freud and Jacob:

We are then taken along the biologistic detour via the genetics of the Period. This is the only section that Freud acknowledged was not yet edited at the death of his daughter—mother of his grandson. These few pages are to be reread in and of themselves in relation to both *The Logic of the Living* [*The Logic of Life*], and that which we had previously accentuated within it: concerning death (immanent or not), sexuality (original or late), protozoa (immortal or not), and the logic of the "supplement," whose ineluctable program we had pointed out. In their principial schemas the two books remain astonishingly contemporary. The new content of scientific advances and of positive discoveries has not, since 1920, displaced the slightest conceptual element in the position of the problems, the kinds of questions, and of the answers or non-answers.[43]

To support Derrida's thesis it is worth noting that Weismann, the biologist Freud refers to when searching for a confirmation of his speculation, is also a key reference for Jacob.[44] In *The Logic of Life*, Jacob acknowledges that it is thanks to Weismann that the numerous theories about the possibility of transmitting acquired characters had been definitively liquidated. For Weismann, living substance consists of two different kinds of cells:

somatic cells, which are in charge of the construction of the organism's body, and germinal cells, which are in charge of reproduction. The former are produced by the singular individual, can be modified from the outside, and are mortal; the second belong to the species, perpetuate from generation to generation, and are immutable and immortal. Attributing hereditary transmission to the second kind of cell, Weismann rejects the transmission of acquired characters and prepares the ground for the formulation of the genetic program as the condition of hereditariness:

> For Weismann, however, the environment can no longer direct heredity. For him, the germ line is beyond the reach of any variation that might occur in individuals of the species. None of the supposed transmissions of acquired characters stands up to analysis. None of the organisms that are mutilated generation after generation produces mutilated descendants. Even when the tails of mice are systematically cut off at birth for five generations, hundreds of little mice continue to be born with normal tails of the same average length as their antecedents. Heredity is proof against any individual whims, any influences, desires or incidents. It resides in the arrangement of matter. According to Weismann, "The essence of heredity is the transmission of a nuclear substance of specific molecular structure." Only changes in this substance, or "oscillations," are able to cause lasting changes in living beings. The whole mechanism of heredity, variation and evolution rests, not on the perpetuation of acquired characters through successive generations, but on the nature of a molecular structure.[45]

Freud recognizes in the distinction formulated by Weismann a remarkable analogy with the distinction between the death and life drives. For this reason, he reconsiders the latter as the dynamic translation of the distinction elaborated by Weismann from a morphological perspective:

> What strikes us in this is the unexpected analogy with our own view, which was arrived at along such a different path. Weismann, regarding living substance morphologically, sees in it one portion which is destined to die—the soma, the body apart from the substance concerned with sex inheritance—and an immortal portion—the germ-plasm, which is concerned with the survival of

the species, with reproduction. We, on the other hand, dealing not with the living substance but with the forces operating in it, have been led to distinguish two kinds of instincts [*Trieben*]: those which seek to lead what is living to death, and others, the sexual instincts [*Sexualtriebe*], which are perpetually attempting and achieving a renewal of life. This sounds like a dynamic corollary to Weismann's morphological theory.[46]

However, there is a noteworthy limit to this analogy. For Weismann, the distinction between somatic cells and germinal cells occurs only in multicellular organisms; monocellular organisms are substantially undifferentiated. Hence, monocellular organisms reproduce themselves by fissure, the one identical to the other, and thus are potentially immortal, whereas death becomes a necessary adaptation for those multicellular organisms that, thanks to the differentiation between somatic and germinal cells, can die as the species perpetuates itself through germinal cells and thus through sexual reproduction. Therefore, for Weismann, as well as for Jacob, death and sexual reproduction are supplementary factors that occur only recently in the evolution of life, in order to assure the essential and originary function of reproduction. The more general framework of Weismann's theory does not merely represent a limit for the analogy proposed by Freud but also contradicts the latter's presuppositions: if for Freud the living substance is animated and structured by the death drives, conversely, for Weismann, it acts in view of reproduction and thus of the perpetuation of life. Freud seems to acknowledge this point:

> Death is rather a matter of expediency, a manifestation of adaptation to the external conditions of life; for, when once the cells of the body have been divided into soma and germ-plasm, an unlimited duration of individual life would become a quite pointless luxury. When the differentiation had been made in the multicellular organisms, death became possible and expedient. Since then, the soma of the higher organisms has died at fixed periods for internal reasons, while the protista have remained immortal. It is not the case, on the other hand, that reproduction was only introduced at the same time as death. On the contrary, it is a primal characteristic of living matter, like growth (from which it originated), and life has been continuous from its first beginning upon earth. It will be seen at once that to

concede in this way that higher organisms have a natural death is of very little help to us. For if death is a *late* acquisition of organisms, then there can be no question of there having been death instincts [*Todestriebe*] from the very beginning of life on this earth.[47]

However, for Freud, the contradiction is only apparent as it concerns the phenomenal manifestation of death: the fact that monocellular organisms look immortal does not mean that the tendency to death, which manifests itself explicitly in multicellular organisms, is not present, latent in them as well. The analogy is safe, and so is the oppositional dualism it presupposes: death drives/life drives, germinal cells/somatic cells. Freud writes:

The instinctual forces [*Triebkräfte*] which seek to conduct life into death may also be operating in protozoa from the first, and yet their effects may be so completely concealed by the life-preserving forces that it may be very hard to find any direct evidence their presence. We have seen, moreover, that the observations made by biologists allow us to assume that internal processes of this kind leading to death do occur also in protista. But even if protista turned out to be immortal in Weismann's sense, his assertion that death is a late acquisition would apply only to its *manifest* phenomena and would not make impossible the assumption of processes *tending* towards it. . . . The striking similarity between Weismann's distinction of soma and germ-plasm and our separation of the death instincts [*Todestriebe*] from the life instincts [*Lebenstrieben*] persists and retains its significance.[48]

Not only is the analogy safe, but it can be further translated into Schopenhauer's philosophical dualism, in which death is the end of life while sexual desire embodies the will to life,[49] and thus the way can be paved for its further translation as the psychoanalytic theory of libido:

Accordingly, we might attempt to apply the libido theory which has been arrived at in psycho-analysis to the mutual relationship of cells. We might suppose that the life instincts or sexual instincts [*Lebens-oder Sexualtriebe*] which are active in each cell take the other cells as their object, that they partly neutralize the death instincts [*Todestriebe*] (that is, the processes set up by them) in those cells and thus preserve their life; while the other cells do the same for *them*, and still others sacrifice themselves in the performance of this libidinal function.⁵⁰

I draw attention to this passage as we may find here an anticipation, even if at an embryonal stage, of the theory of *apoptosis* or cellular suicide. Formulated only recently, at the end of the twentieth century, this theory explains the cellular differentiation required by the construction of a living organism as a result of the phenomena of cellular death induced by the cells of the organism itself. The expression "cellular suicide" refers to the hypothesis that the death of the cell occurs as a response to signals emitted by other cells of the same organism. I discuss this theory in greater detail in the next chapter, where I aim to demonstrate that the theory of autoimmunity formalized by Derrida since the 1990s essentially depends on the theory of *apoptosis*. Derrida's remarks on Freud's passage seem to legitimate this insight:

> At first he seems to turn this model toward a politico-psychoanalytic metaphor: the vital association of cells in order to preserve the life of the organism. The State, or the multicellular society, guards life beyond the death of any given subject. The primitive *socius*, the original, "natural" contract: copulation serves both reproduction and the rejuvenation of the other cells. One *could* at this point play upon the transferential metaphor, transfer the transference, and compare, *übertragen* says Freud, the psychoanalytic theory of libido with these bio-political cells. Present in every cell, the two drives (life, death) partially neutralize the effects of the death drive in the other cells which they are keeping alive, occasionally pushing the thing to the sacrifice of themselves. This sacrifice, of course, would be coordinated with the great reckoning, the great economy of the inheritance.⁵¹

As I postpone my demonstration till the next chapter, let me go back to Freud for the moment. The extension of the theory of libido to cellular behavior and thus the hypothesis that the life of the living is oriented by the death drives and develops through the struggle with the life drives has no scientific support. Freud is forced to rely on the myth of the androgynous, as it is demonstrated by Aristophanes in Plato's *Symposium*.⁵² The correspondences this myth has with the analogous myth preserved in the *Upaniṣads* leads Freud to attribute a certain credit to the one narrated by Plato through Aristophanes. In any case, Freud ends his speculation at this point, before reaching a satisfactory conclusion about the hypothesis of the death drives. Rather, he seems to return to the point of departure, the initial state, that is to say, to the pleasure principle. Finally, the only support for the hypothesis that the death drives are a regressive pressure that induces the living to return to the state of inertia of non-living matter is precisely the pleasure principle. Since the beginning, I note, Derrida has highlighted the teleological orientation of the pleasure principle and thus of its determination in view of the *telos* of the death drives, beyond Freud's rhetoric, which seems to grope around throughout the text among speculative hypotheses that are first advanced, then abandoned, then taken up again and translated in various directions without ever coming to any determinate conclusion:

> The dominating tendency of mental life, and perhaps of nervous life in general, is the effort to reduce, to keep constant or to remove internal tension due to stimuli (the "Nirvana principle" to borrow a term from Barbara Low)—a tendency which finds expression in the pleasure principle; and our recognition of that fact is one of our strongest reasons for believing in the existence of death instincts [*Todestrieben*].[53]

The last chapter of *Beyond the Pleasure Principle* comes back to the point of departure in order to reconstruct in summary the path the text has followed, and yet, for Derrida, it is not a mere recapitulation.

From the Law of the Proper to the Drive for Power and Beyond

Freud returns to his initial problem as if he had said nothing about the death drives' anticipation in relation to defining the pleasure principle's tendency, that is to say, "reducing, keeping constant, eliminating the internal tension produced by stimuli." Once the independence of the repetition drives from the pleasure principle is confirmed, what remains is to establish the nature of the relationship between the drives and the principle:

> This characteristic would be shared by all the component instincts [*Partialtriebe*] and in their case would aim at returning once more to a particular stage in the course of development. These are

matters over which the pleasure principle has as yet no control; but it does not follow that any of them are necessarily opposed to it, and we have still to solve the problem of the relation of the instinctual [*triebhaften*] processes of repetition to the dominance [*Herrschaft*] of the pleasure principle.[54]

He recalls and generalizes the binding function that was found to be at work in repetition compulsion by identifying it with a genetico-structural condition of the psychic apparatus:

We have found that one of the earliest and most important functions of the mental apparatus is to bind the instinctual impulses [*Triebregungen*] which impinge on it, to replace [*zu ersetzen*] the primary process prevailing in them by the secondary process and convert their freely mobile cathectic energy into a mainly quiescent (tonic) cathexis. While this transformation is taking place no attention can be paid to the development of unpleasure; but this does not imply the suspension of the pleasure principle. On the contrary, the transformation occurs on *behalf* of the pleasure principle; the binding is a preparatory act which introduces and assures the dominance of the pleasure principle.[55]

This is a decisive step for Derrida. Here he finds the resources required to resolve the problem he had taken as his own point of departure since Edmund Husserl's "Origin of Geometry," searching the conditions of the retentional trace and thus of the processes of idealization depending on them, to begin with the genesis of the living from which consciousness emerges. The binding function described by Freud among primary (originary, unconscious, spontaneous, and thus "natural") and secondary (conscious, symbolic, ideal, and thus "cultural") processes allows Derrida to describe those conditions and their articulation in the transition to consciousness: binding the primary drive processes, which originarily master psychic life without opposition, with the secondary drive processes that characterize the conscious condition means replacing the former with relatively stable representatives that are able to stabilize the free flow of energy that characterizes the primary processes and, at the same time, to represent that flow:

This "function" (*Funktion*) is the *Binden*, the operation which consists in binding, enmeshing, tying up, garroting, tightening, handing. But what? Well, that which is as original as this function of stricture, to wit, the forces and excitations of the drives, the X about which one does not know what it is before it is handed, precisely, *and* represented by representatives. For this early and decisive function consists of binding *and* of replacing: to bind is immediately to supplement, to substitute, and therefore to represent, to replace, to put an *Ersatz* in the place of that which the stricture inhibits or forbids. To bind, therefore, is also *to detach*, to detach a representative, to send it on a mission, to liberate a missive in order to fulfill, at the destination, the destiny of what it represents. A *post* effect.[56]

Through the genesis of the substitutive binding of the primary processes, we detect the genesis of the trace as the iterable trace, and, at the same time, we account for its structure as well as for the energetic dynamic that necessarily constitutes it as such and thus as an articulation or effect of différance (differential/differing). Iterability (repetition) assures the stabilization of the free flow of energy that the trace represents, but, to this end, that is, in order to refer to the flow, the trace must necessarily be different. For this reason, the trace cannot be the trace of an originary presence (the primary processes and the free flow of energy that constitutes them are necessarily unstable and unconscious, therefore, structurally non-present) but the elaboration of a secondary, supplementary presence through which it is possible to refer to the differences of force (free energy/bound energy, investment energy/energy of counter-investment) that generated it but whose presence it is impossible to make manifest. Therefore, the trace takes place only through a web of references in which it is inscribed and not in and by itself, namely, as the supposed trace of an already supposed primordial, intuitive presence (to use the language of phenomenology). Furthermore, given the law of general textuality that structures the living and its products as weaves of iterable traces, the system of references in which the trace is elaborated and inscribed cannot be structurally saturated. Therefore, however stable the formation of the trace is, even if it is always liable to further alterations/iterations, the bond that it establishes and represents can be resolved and reconstituted again in a different way, according to different conditions and forces that intervene in the elaboration of the system:

> In the same statement, describing one and the same operation, one and the same function, Freud says that it consists of bind-

ing (*Binden*) the primary process (*pp*) and of replacing (*ersetzen*) the *pp* which has mastery (*herrschenden*) over the life of the drives with the secondary process: displacement, replacement of mastery, stricture as supplementary detachment. The secondary is the supplementary *sending* (*envoi*). It transforms freely mobile cathectic energy into immobile cathectic energy, it *posits* and *posts*. Now here is a thesis. The immobilized cathexis becomes more tonic. The notion of tonicity regularly finds itself associated with the effect of binding, which thus signifies both elasticity and tension. Which consolidates the legitimacy of the translation of *hinden* by *to band* [*bander*]. And, taking into account the supplementary relays that I have just recalled, to post: to band. Postal: binding.[57]

In this case, the "*bindinal* economy"[58] described by Derrida in these pages of *The Post Card* must be understood as the economy to which he had already referred in "Force and Signification."[59] It is able to account for the relationship between force and form or between genesis and structure as a differential relationship and not an opposition, that is, not according to a classic metaphysical scheme in which, as we saw, both phenomenology and structuralism remain trapped. But let me go back to the last pages of *Beyond the Pleasure Principle*. For Freud, the binding function is a primordial origin of the structuration of the psychic apparatus and thus is indifferent to pleasure and displeasure; it precedes the institution of the pleasure principle and makes it possible. However, for him, not only does the binding function precede the pleasure principle as its condition of possibility, but it also opens the way, prepares the ground (the psychic apparatus) for the exercise of the pleasure principle's sovereignty. Derrida, in turn, detects the almost imperceptible shift that Freud produces between preceding and preparing. In so doing, he highlights the conceptual stretching by means of which the psychoanalyst subordinates the binding function, which is itself indifferent to pleasure and unpleasure, to the *telos* that since the beginning had oriented the description of the pleasure principle, and thus to the confirmation of the hypothesis of the death drives and the general tendency of the living to restoring the state of inertia proper to inorganic matter:

> [I]f, as such, binding is not yet accompanied by either pleasure or unpleasure, if at least it can be isolated from them, where is this *preparatory act* to be situated? What does to prepare signify in

this case? What about this *pre*? It is simultaneously, in this *lapse* or *capsule*, indifferent to pleasure and to unpleasure *and* rather interested, inspired, called upon, by the PP, since it announces the PP in its turn and *makes room* for it. It precedes and prefigures it. Of the two modes of the *pre*, only the latter one seems teleological. The first one seems indifferent. How to adjust the *telos* to indifference, the ends of one to the ends of the other?[60]

Indeed, Freud submits the binding function, that is, the general structural condition of the psychic apparatus, to the pleasure principle understood as a tendency that is in turn subordinated to a more general function, that of bringing the psychic apparatus to a state of acquiescence free from excitation, which, again, is subordinated to the death drives:

> The pleasure principle, then, is a tendency operating in the service of a function whose business it is to free the mental apparatus entirely from excitation or to keep the amount of excitation in it constant or to keep it as low as possible. We cannot yet decide with certainty in favour of any of these ways of putting it; but it is clear that the function thus described would be concerned with the most universal endeavour of all living substance—namely to return to the quiescence of the inorganic world.[61]

Hence, for Derrida, the pleasure principle consists in the middle term that makes possible the transition from the binding function to the *telos* of the return to the inorganic, and thus the development of Freud's argument reproduces a dialectic-speculative movement of a Hegelian kind, even if it is inverted:

> At this point the PP would not be a function but a tendency in the service of this general function. But it would have another function (binding) in its service. The general functioning would move from one function to the other, from the function of the *Binden* to the function in its most general form (return to the inorganic and Nirvana) via the intermediary or place of passage, the *step* [*pas*] of a *tendency*, *to wit* the PP. The *pas de PP* between two functions or two forms of general functioning.[62]

However, to hold together the process and its *telos*, Freud formulates a paradoxical, if not intimately contradictory, conclusion. The pleasure principle is instituted as it limits pleasure; it unfolds its mastery over itself, and its authority increases to the extent that the possibility of pleasure, unchained and limitless in the primary processes, is bound and reduced in the secondary processes:

> We thus reach what is at bottom no very simple conclusion, namely that at the beginning of mental life the struggle for pleasure was far more intense than later but not so unrestricted: it had to submit to frequent interruptions. In later times the dominance [*Herrschaft*] of the pleasure principle is very much more secure, but it itself has no more escaped the process of taming than the other instincts [*Triebe*] in general.[63]

For Derrida, the pleasure principle is instituted as self-mastery insofar as it subordinates itself to the binding function that makes mastery possible by restricting the tendency to pleasure and limiting mastery itself. This works also for the secondary processes in which the reality principle is at play: as we saw, the pleasure principle exercises its mastery by differing (from) itself as the reality principle, to which it must be bound and subordinated in order to be itself; if the pleasure principle can be itself only through this indefinite differential/differing reference (*Umweg*), then its self-mastery will never be absolute but always relative and dependent on the possibility of being bound to itself in this differential/differing movement:

> If it is to assure its mastery, the principle *of* pleasure therefore first must do so *over* pleasure and at the expense *of* pleasure. Thus it becomes the prince *of* pleasure, the prince whose pleasure is the conquered, chained, bound, restricted, tired subject. The game is necessarily played on two boards. Pleasure loses in *measure* itself: in which it brings its principle to triumph. It loses on every turn, it wins on every turn *by measure* of its being there before being there, as soon as it prepares itself for its presence, by measure of its still being there when it reserves itself in order to produce itself, invading everything beyond itself. It wins on every turn, it loses on every turn *by measure*: its unleashed intensity would destroy it immediately if it did not submit itself to the

moderating stricture, to measure itself. Death threat: no more principle of pleasure therefore no more *différance* that *modifies* it into a reality principle. What is called reality is nothing outside this law of *différance*. Reality is an effect of this law. Stricture produces pleasure by binding it.[64]

Derrida observes that the binding function, namely, the function of stricture, is the irreducible condition of possibility of the constitution of the psychic system understood as a living system, but also, on the basis of the law of general textuality as the genetico-structural condition of the living, of any unitary formation that the living produces:

> The force of stricture, the capacity to *bind itself*, remains in relation to *what there is to bind* (what gives something and gives itself to be bound), the power binding the binding to the bindable. . . . Of course what we are saying here is already valid for what we are calling the "set" itself. If this word is to refer to a "unity" which rigorously is neither that of the subject nor that of consciousness, the unconscious, the person, the soul and/or the body, the socius or a "system" in general, then it is indeed necessary that the set as such *bind itself* to itself in order to constitute itself as such. Every belt together, even if its modality is not limited to any of those we have just placed in a series, begins by *binding-itself*, by a binding-itself in a differantial relation to itself. It thereby sends and posts itself. Destines itself. Which does not mean: it arrives.[65]

Derrida suggests that, for Freud, this differential/differing function of stricture, recognized as the irreducible condition of any living unity, whole, or system, is in turn bound to the exercise of a mastery that would be more originary and pervasive than the other masteries taken into consideration so far (pleasure principle/reality principle) and, ultimately, would preside over them: "There would be, bound to stricture, and by means of it, a notion of mastery which would be neither of life nor of death. It would be even less what is at stake in a struggle of consciousness or a struggle for recognition. And sexuality would no longer determine it in the last analysis."[66] Derrida highlights Freud's formulation of this drive for mastery in *Beyond the Pleasure Principle*, even if it only consists of a quick reference, justified by the possible interpretation of the deep motivations that would bring his grandson Ernst to play with the "wooden reel," namely, to throw the latter and retrieve it by pulling the thread it is tied to it. The game could be explained, Freud suggests, as the manifestation of a *"Bemächtigungstrieb,"* a drive for mastery, which induces the child to master the unpleasant experience related to the temporary absence of the mother through the voluntary repetition of the same experience:

> No certain decision can be reached from the analysis of a single case like this. On an unprejudiced view one gets an impression that the child turned his experience into a game from another motive. At the outset he was in a *passive* situation—he was overpowered by the experience; but, by repeating it, unpleasurable though it was, as a game, he took on an *active* part. These efforts might be put down to an instinct for mastery [*Bemächtigungstrieb*] that was acting independently of whether the memory was in itself pleasurable or not.[67]

Here Derrida finds the condition of possibility immanent in all drives: the self's drive to master itself, to appropriate itself through the appropriation of the other that the self needs in order to relate to itself and thus to seize itself. Evidently we are before a reinscription/translation of the law of the proper into a "bindinal" economy:

> In question, then, is a simple allusion, but what the allusion designates calls upon the singularity of a drive that would not permit itself to be reduced to any other. And it interests us all the more in that, being irreducible to any other, it seems to take part in all the others, in the extent to which the entire economy of the PP and its beyond is governed by relations of "mastery." One can envisage, then, a quasi-transcendental privilege of this drive for mastery, drive for power, or drive for domination [*emprise*]. The latter denomination seems preferable: it marks more clearly the relation to the other, even in domination *over oneself*.... The drive to dominate must also be the drive's *relation to itself*: there is no drive not driven to bind itself to itself and to assure itself of mastery over itself as a drive. Whence the transcendental tautology of the drive to dominate: it is the drive as drive, the drive of the drive, the drivenness of the drive. Again, it is a question of a relation to oneself as a

relation to the other, the auto-affection of a *fort:da* which gives, takes, sends and destines itself, distances and approaches itself by its own step, the other's.[68]

Although this notion appears already in *Three Essays on the Theory of Sexuality* and returns occasionally in later texts, Freud never explicitly thematizes the drive for power as such, least of all from the perspective adopted by Derrida. However, Derrida draws attention to another passage from *Beyond the Pleasure Principle* in which Freud aims to account for primary sadism as a possible manifestation of the death drive (chapter VI):

> From the very first we recognized the presence of a sadistic component in the sexual instincts [*Sexualtriebes*]. As we know, it can make itself independent and can, in the form of a perversion, dominate [*beherreschen*] an individual's entire sexual activity. It also emerges as a predominant component [*als dominierender Partialtrieb*] in one of the "pregenital organizations," as I have named them. But how can the sadistic instinct [*Trieb*], whose aim it is to injure the object, be derived from Eros, the preserver of life? Is it not plausible to suppose that this sadism is in fact a death instinct [*Todestrieb*] which, under the influence of the narcissistic libido, has been forced away from the ego and has consequently only emerged in relation to the object? It now enters the service of the sexual function. During the oral stage of organization of the libido, the act of obtaining erotic mastery [*Liebesbemächtigung*] over an object coincides with that object's destruction; later, the sadistic instinct [*Trieb*] separates off, and finally, at the stage of the genital primacy, it takes on, for the purpose of reproduction, the function of overpowering [*zu bewältigen*] the sexual object to the extent necessary for carrying out the sexual act.[69]

For Derrida, these elements are enough to affirm that, even if unconsciously, the drive for power dominates the organization of Freud's entire discourse:

> Now, if such a drive for power exists, if it sees itself attributed a specificity, then it indeed has to be admitted that it plays a

very original role in the most "meta-conceptual," "metalinguistic," precisely the most "dominant" organization of Freudian discourse. For it is indeed within the code of power, and this is not only metaphorical, that the problematic is lodged. It is always a question of knowing who is the "master," who "dominates," who has "authority," to what point the PP exercises power, how a drive can become independent of it or precede it, what are the relations of service between the PP and the rest, what we have called the prince and his subjects, etc.[70]

In particular, for Derrida, the dynamic of sadism, the possibility that it subordinates all the other drives, is due to this more originary and general drive for power. It is a possible articulation or manifestation of this drive, and so are the death drive and the pleasure principle to the extent that they aim at hegemony. The drive for power dominates all the other drives and allows us to account for them:

> There is a society of drives, whether or not they are communally possible, and in the passage to which we have just referred (chapter VI), the dynamics of sadism are dynamics of power, dynamics of dynasty: a component drive must come to dominate the entirety of the body driven, and must subject this body to its regime: and if this succeeds, it is with the aim of exercising the violence of its domination over the object. And if this desire to dominate is exercised within as well as without, if it defines the relation to oneself as the relation to the other of the drives, if it has an "original" root, then the drive for power *can* no longer be derived. Nor can postal power. In its auto-heterology the drive for postal power is more originary than the PP and independent of it. But it equally remains the only one to permit the definition of a death drive, and for example an original sadism. In other words, the motif of power is more originary and more general than the PP, is independent of it, is its beyond. But it is not to be confused with the death drive or the repetition compulsion, it gives us with what to describe them, and in respect to them, as well as to a "mastery" of the PP, it plays the role of transcendental predicate. Beyond the pleasure principle—power. That is, posts.[71]

However, the fact that the drive for power consists of the condition of possibility of the other drives, that it dominates them from a transcendental position, does not mean that the subordinated drives accomplish in full its end or aim. In order to master itself, to appropriate itself, to realize itself as such, the drive for power must dominate the other drives and thus must differ (from) itself in the others. Therefore, the accomplishment of the tendency to absolute power is necessarily differed/differentiated. For this reason, power can only be a relatively stable effect of the general dynamic of différance and the stricture in which it is inscribed:

> But even so, we will not say, despite the transcendental function to which we have just alluded, beyond the death drive—power—or posts. For it is equally the case that everything described under the heading of the death drive or the repetition compulsion: although proceeding from a drive for power, and borrowing all its descriptive traits from this drive, no less overflows power. This is simultaneously the reason and the failure, the origin and the limit of power. There is power only if there is a principle or a principle of the principle. The transcendental or meta-conceptual function belongs to the order of power. Thus there is only *différance* of power. Whence the posts.[72]

Even if it is always urgent and pervasive, the drive for power never accomplishes itself as such or as an absolute power; it has always to negotiate its hegemony with other forces in the field or, more simply, with the other over which it wants to exercise its hegemony. Therefore, the drive for power can be wrestled away from its potentially devastating effects: sadism, cruelty, and death. Perhaps the self, too—the constitution of the self—could be wrestled away from the drive for power in view of another binding (individual or collective, as it would be either living or elaborated by the living). As we have seen, for Derrida, the possibility of binding or stricture consists in the irreducible condition of possibility of the constitution of a whole, a system (whether individual or collective, natural or cultural), and thus also of the exercise of the drive for power. However, he noted that the function of binding or stricture is bound to the drive for power but does not depend on it, which means that this binding can be resolved and the relative function can be bound otherwise. It can be bound to something else; it can bind the drive for power to itself otherwise and thus make room for relations of power that are different from those that imply the violent subordination of the other, of the self (and) of the other, up to death. It is from this point, that is, from the question of the drive for power as the condition of the constitution of the self as a living form, that we must follow Derrida's search for a beyond of the drive for power, which will lead him to thematize the questions of the death penalty and, later, of sovereignty. This connection, which is already at stake in the conference on *Beyond the Power Principle*, given in New York in 1985,[73] and later in *Being Just with Freud* (1992), is made explicit in "Psychoanalysis Searches the States of Its Soul" (2000), included in *Without Alibi*:

> At the more distant horizon of these questions would loom the necessity to situate, along with the psychoanalytic theme of sovereignty or mastery (*Herrschaft*, *Bemächtigung*), which is so present in *Beyond the Pleasure Principle*, at least in the form of a political metaphor, the theme of a *Bemächtigungstrieb*, a drive for ascendancy, for power, or for possession. I tried to show elsewhere, in a long *Post Card*, how the word and concept of *Bemächtigung*, however discreet they are and however underanalyzed by Freud's readers, are present beginning in the *Three Essays* and play in *Beyond* a decisive role, beyond or on this side of the principles, precisely, as principial drive, if I can say that, notably in love/hate ambivalence and the unleashing of cruelty that calls up the hypothesis of originary sadism. Indissociable from that of *Bewältigung* (exercise of power, ascendancy, or possession, movement of appropriation, etc.), there would thus be the concept of a *drive for power*—that is to say, of the capacitation [*de l'habilitation*], of the "I can" or "I may," and in particular of the performative power that organizes, via some sworn faith, the whole order of what Lacan called the symbolic.[74]

In this context, to conjure away the devastating effects of the cruel drive for power, Derrida appeals to the necessity of reinscribing it in the economy of the detour (*Umweg*), namely, of the différance that structures the living:

> Freud believes in the ineradicable existence of drives of hatred and destruction. Making very frequent use of the words "cruelty," "aggression drive," "hatred drive," and "death drive," he denounces an illusion: that of an eradication of the cruelty drives

and the drives for power and sovereignty. "What it is necessary to cultivate (for it is necessary that an "it is necessary," and thus the tie of an ethical, juridical, political obligation, take shape) is a differential transaction, an economy of detour and difference, the strategy, one can even say the method (for it is a question here of path, path breaking, and road), of indirect progress: an indirect, always indirect way of combating the cruelty drive. The word "indirect" is articulated like the pivot of this progressivism without illusion.[75]

I note that, in "Psychoanalysis Searches for the States of Its Soul," Derrida translates "*Bemächtigungstrieb*" as "drive for power or for sovereign mastery"[76] and conceives of the death penalty as the institution through which cruelty is inscribed at the heart of political sovereignty:

It is next a matter, more precisely, of the indissociable tie between cruelty and state sovereignty, state violence, the state that, far from combating violence, monopolizes it. A few years later, this will be Benjamin's theme in *Critique of Violence*, around which I elaborated a few propositions on law (or right) and justice in "Force of Law." This monopoly on violence is of a piece with the motif of sovereignty. It is also what will always have grounded the death penalty, the right of the state, the right of the sovereign to punish by death.[77]

Therefore, it is not by chance that we find the discussion of the "*Bemächtigungstrieb*," the "drive for power or for sovereign mastery," at a key moment in *The Beast and the Sovereign I*:

[W]hat I am here designating as transfer of sovereignty clearly situates the essential features of the problem. If most often what is at stake in politics and wherever else a drive to power is exercised (*Bemächtigungstrieb*, as Freud calls it, before or beyond the other drives and the death drive), a drive to power that orders even the drive to see and to know, the scopic and epistemophilic drive—(if most often what is at stake in politics and wherever else a drive to power is exercised) is not only an alternative between sovereignty and nonsovereignty but also a *struggle* for sovereignty, transfers and displacements or even divisions of sovereignty, then one must begin not from the pure concept of sovereignty but from concepts such as drive, transference, transition, translation, passage, division. Which also means inheritance, transmission, and along with that the division, distribution, and therefore the economy of sovereignty. . . . [rather than on sovereignty itself], it is on these properly mediate words and concepts, impure like middles or mixtures (words and concepts such as transfer, translation, transition, tradition, inheritance, economic distribution, etc.) that we must bring the charge of the question and of decisions that are always median, medial transactions, negotiations in a relation of force between drives to power that are essentially divisible.[78]

It is within this horizon that Derrida's more recent works on the death penalty, sovereignty, and democracy to come should be inscribed. To understand their strictly political bearing, it is necessary to address the problem of sovereignty since its elaboration as an articulation of the differential/differing dynamic of *life-death* that structures the psychic system, namely, the self, as well as the systems and wholes that the latter produces, and thus also the political, sovereignty, and so forth, as an emergence of the living system. This works also for autoimmunity, a concept to which Derrida turns more and more often, after the 1990s in particular, to account for the essential dynamics that structure democracy. In this case too, as we shall see in the next chapter, we must retrace the specific significance of autoimmunity back to the differential/differing dynamic of *life-death* that constitutes the living; in other words, we must go back again to the biological roots of this new import.

In the final chapter, dedicated to survival, I try to trace the path through which, beyond Freud, Derrida dissociates the function of binding from the drive for power (which, for Freud, is inextricably tied to the former) and thus rearticulates it with différance, at the heart of the living (*life/death*), in order to dissociate the genesis and structure of the living self, and thus the emergence of the psychic system, from the hold of the drive for power or for sovereign mastery.

VI

Beyond Life Death: Autoimmunity

After *Specters of Marx* (1993), Derrida turned more and more frequently to the notion of autoimmunity, especially in relation to religion, community, democracy, and sovereignty. However, he did not devote a specific and detailed analysis to the notion, and its occurrences, disseminated throughout Derrida's corpus, sound quite enigmatic, often ambiguous, and rather contradictory. In *Faith and Knowledge* (1994–1996), which Derrida himself indicates as the source for his treatment of autoimmunity, we find an explicit reference to biology:

> It is especially in the domain of biology that the lexical resources of immunity have developed their authority. The immunitary reaction protects the "indemn-ity" of the body proper in producing antibodies against foreign antigens. As for the process of auto-immunization, which interests us particularly here, it consists for a living organism, as is well known and in short, of protecting itself against its self-protection by destroying its own immune system. As the phenomenon of these antibodies is extended to a broader zone of pathology and as one resorts increasingly to the positive virtues of immuno-depressants destined to limit the mechanisms of rejection and to facilitate the tolerance of certain organ transplants, we feel ourselves authorized to speak of a sort of a general logic of auto-immunization.[1]

To uncover the reasons for and the significance of Derrida's recourse to an autoimmunitarian lexicon, the formalization of the latter into a general logic, and thus the structural and structuring function it plays within deconstruction, we must go back to the biological sources from which this lexicon is explicitly imported. As we shall see, on the one hand, it is only from this perspective that we can understand autoimmunity as a consistent development of the seminar *La vie la mort* and thus as a further articulation of différance (life/death) as the genetico-structural condition of the living. On the other hand, consequently, it is only from this perspective that we can take account of one of the traits that Derrida constantly associates with autoimmunity and the processes of autoimmunization, and thus we can avoid the risk of misleading interpretations. In other words, the self-destructive character, the suicidal tension that, for Derrida, structures autoimmunity and the processes of autoimmunization, could be understood as a merely negative consequence of the exposure to the other that necessarily conditions the life of the living and without which the living could not be what it is.[2] As we shall see, only the reference to biological sources allows us to find in the self-destructive or suicidal character of the autoimmunitarian processes the irreducible condition that structures the life (death) of the living and not an impending, autoimmune threat. Autoimmunity is not a possible, negative consequence of the exposure to the other but the genetico-structural condition of this exposure: it explains the structure and function of this exposure from its biological genesis. However, if we understand self-destruction and suicide as an irreducible and structuring condition of the life of the living, without any reference to the seminar *La vie la mort* and to the biological context that legitimates the recourse to these notions, we would come to a paradoxical, if not merely contradictory, interpretation. We should explain why self-destruction and suicide—the pure and simple negation of life—would be for Derrida the structuring condition of life.

There is no doubt that Derrida's reference to HIV as an example of autoimmunitarian pathology in *Faith and Knowledge* produced confusion. In a passage between parentheses within a note, Derrida refers to the papal encyclical dedicated to the sacredness of life as follows: "the last encyclical *Evangelium Vitae*, against abortion and euthanasia, for the sacredness or holiness of a life that is safe and sound—unscathed, *heilig*, holy—for its reproduction in conjugal love—sole immunity admitted, with priestly celibacy, against human immuno-deficiency virus (HIV)."[3] However, the confusion cannot be ascribed to Derrida but to the ignorance of the biological context from which these notions are drawn. As we shall see, autoimmunity is a structural function of the living organism; it belongs to its ordinary constitution and, in particular, presides over the construction and functioning of the immunitarian system and thus the constitution of the identity of the living individual. Conversely, autoimmune diseases, such as HIV, are pathologies that affect the immunitarian system by modifying their ordinary functioning. Derrida does not construct his general autoimmunitarian logic on the basis of autoimmune pathologies, although he is certainly interested in the pathological modifications that the ordinary autoimmunitarian processes, which are essential to the life of the living, can undergo. But before developing a detailed analysis of Derrida's use of the autoimmunitarian lexicon and thus searching for the biological sources from which it is borrowed, I highlight some traces that allows us to establish with some precision the continuity between these problematics and the seminar *La vie la mort*.

First of all, Derrida retraces the autoimmunitarian process back to the question of the general textuality that, as we know, structures the life of the living from its genetic constitution: "The living organism destroys the conditions of its own protection. Such auto-immunization is a terrifying biological possibility: a body destroys its proper defenses or organizes in itself (and this is, in an extended sense, a question of genetic writing and reading [omitted in the English translation]) the destructive forces that will attack its immunitary reactions."[4] In *The Animal That Therefore I Am*, the articulation between autoimmunity and the textual structure of the living is even more evident:

> Autobiography, the writing of the self as living, the trace of the living for itself, being for itself, the auto-affection or auto-infection as memory or archive of the living, would be an immunizing movement (a movement of safety, of salvage and salvation of the safe, the holy, the immune, the indemnified, of virginal and intact nudity), but an immunizing movement that is always threatened with becoming auto-immunizing, like every *autos*, every ipseity, every automatic, automobile, autonomous, auto-referential movement.[5]

The autoimmunitarian movement is inscribed in the textual dynamic that structures the living; it is an irreducible consequence of that dynamic. If the living must identify itself with the texture of differential traces that structure the living itself as such, and if it must do this in order to be itself, then the identity of the living will never be pure, immediate, closed on itself, and immune from any relationship with the other, because the trace, the elementary structure of the living, which implies in itself the referral to something other than itself, is anything but this referral. Therefore, if the process of self-constitution of individual identity is a process of immunization with respect to alterity in general, this process must necessarily ("*toujours*") imply in itself an autoimmunitarian phase, so long as a self that is absolutely immune to the other in general, an identity that is hermetically closed to the other in general, could not be the identity of a living self. The autoimmunitarian is a threat for the self, for the identity that aims toward an absolute autoimmunity, but it is a structural condition for the living self that organizes itself according to the dynamic of differential trace.

Rogues (2003) is the text where the autoimmunitarian plays the key role of describing the aporetic or undecidable conditions that structure democracy as well as the pathologies that affect Reason for Husserl. This text also allows us to grasp the role of the autoimmunitarian in the differential articulation of the constitution of the living self and thus in the emergence of consciousness, according to the coordinates established through the reading of Freud in the seminar *La vie la mort*:

> What psychoanalysts call more or less complacently the unconscious remains, it seems to me, one of the privileged sources, one of the vitally mortal and mortally vital reserves or resources, for this implacable law of the self-destructive conservation of the "subject" or of egological ipseity. To put it a bit sententiously in the interest of time, without autoimmunity there would be neither psychoanalysis nor what psychoanalysis calls the "unconscious." Not to mention, therefore, the "death drive," the cruelty of "primary sadism and masochism"—or even what we just as complacently call "consciousness."[6]

The autoimmunitarian logic would be an essential resource for psychoanalysis to justify Freud's speculations on the origins and genesis of the psychic system at the level of the biological constitution of the living self. Death drive, primary sadism and masochism, and the cruel drive for power, which for Freud structure the life of the living and thus the functioning of the psychic system, could be understood as expressions of autoimmunitarian bio-logic. As we may recall, I have already highlighted that Freud's extension of the death drives to the behavior of the cell in the construction of the organism could represent, on Derrida's reading, an anticipation of autoimmunitarian logic. However, it seems evident that, beyond the multiple thematic contexts in which the autoimmunitarian lexicon intervenes, the latter is adopted to describe a structural and biological condition of the constitution of the living self, and thus the effects of this condition are propagated beyond the living self throughout the processes of the identitarian constitution elaborated by the living self on a level that is no longer "natural" but "cultural." In *Rogues*, Derrida explicitly affirms that he made use of the autoimmunitarian lexicon to reinscribe the themes of democracy and the political within the horizon of the question of the life of the living and in particular of that to which the title *La vie la mort* responds:

> Why determine in such an ambiguous fashion the threat or the danger, the default or the failure, the running aground or the grounding, but also the salvation, the rescue, and the safeguard, health and security—so many [*comme autant de*] diabolically *autoimmune* assurances, virtually capable not only of destroying themselves in suicidal fashion but of turning a certain death drive against the *autos* itself, against the ipseity that any suicide worthy of its name still presupposes? In order to situate the question of life and of the living being, of life and death, of life-death, at the heart of my remarks [*au coeur de mon propos*].[7]

In another passage from *Rogues*, it becomes evident that for Derrida the autoimmunitarian lexicon can be applied to politics in general and, in particular, to democracy only to the extent that it consists in a biological law that constitutes and affects the life of the living as well as of its "cultural" products. For Derrida this does not mean adhering to a reductionist biologism, but rather drawing on the consequences of the law of general textuality as the genetico-structural condition of the living and on the effects of this law as they are developed in *La vie la mort*:

> Why did I think it necessary in order to formalize this strange and paradoxical revolution to privilege today something that might look like a generalization, without any *external limit*, of a biological or physiological model, namely, autoimmunity? It is not, you might well imagine, out of some excessive biologistic or geneticist proclivity on my part. I began by noting that the circular or rotary movement of the self's return to itself and against itself, in the encounter with itself and countering of itself, would take place, as I understand it, before the separation

of *physis* from its others, such as *tekhne*, *nomos*, and *thesis*. What applies here to *physis*, to *phuein*, applies also to life, understood before any opposition between life (*bios* or *zoe*) and its others (spirit, culture, the symbolic, the specter, or death). In this sense, if autoimmunity is *ph*ysiological, *biological*, or zoological, it precedes or anticipates all these oppositions. My questions concerning "political" autoimmunity thus concerned precisely the relationship between the *politikon*, *physis*, and *bios* or *zoe*, life-death.[8]

As we have seen, the movement of the reproduction of the differential/differing trace that conditions the genesis and structure of the living self is necessarily without limit. Thanks to the structure of the trace as the self's differential reference to itself, the movement of différance that structures the living reproduces itself beyond the living self and thus also conditions the "cultural" production of the living itself, even if it occurs through specific articulations and degrees. Therefore, it also conditions democracy and the political, understood as "cultural" products of the living. Autoimmunity evidently belongs to the same differential/differing movement that structures the constitution of the living self at a biological and "natural" level and propagates without limits beyond that level, that is, throughout its so-called "cultural" products, such as democracy and the political.[9] Finally, we should conceive of autoimmunity as a condition of possibility of the entire process of identitarian constitution, of the constitution of the *autos* in itself, and thus of the phenomena in which we can find those identitarian processes at work, from the biological level of the constitution of the living self to the level of individual and collective consciousness, up to the constitution of the community in general and of democracy in particular. Wherever a process of identitarian constitution is at work, subordinated to the desire of an immediate, autonomous, and independent self-presence, pure and intact, immune to the intrusion of the outside, the other, difference, the foreigner—the desire that has always already pulled Western thought among the irresoluble aporias of metaphysics, as Derrida had shown for years—there we must recognize the deconstructive work of autoimmunity. This is said explicitly in *Rogues*:

> For what I call the autoimmune consists not only in harming or ruining oneself, indeed in destroying one's own protections, and in doing so oneself, committing suicide or threatening to

do so, but, more seriously still, and through this, in threatening the I [*moi*] or the self [*soi*], the *ego* or the *autos,* ipseity itself, compromising the immunity of the *autos* itself: it consists not only in compromising oneself [*s'auto-entamer*] but in compromising the self, the *autos*—and thus ipseity.[10]

Therefore, the autoimmunitarian logic is the condition of possibility of any identitarian constitution, of consciousness as well as of the community. But to detect this logic as such we must find it already at work in the constitution of the living, as the condition of possibility of the life of the living and thus of its successive articulations and manifestations. Even if the passages quoted are quite obscure, it is nonetheless clear that Derrida's turn to the autoimmunitarian lexicon is neither metaphorical nor merely formal but refers instead to the biological conditions of the organization of the living as they are reproduced without limits beyond the living itself according to the law of general textuality. For this reason, it seems to me absolutely indispensable to go back to the biological sources from which Derrida borrows. But before taking this step forward, we must look more closely at what Derrida means by "autoimmunity" and "autoimmunitarian."

The first occurrence of the term can be found in *Specters of Marx* (1993), where there are also numerous occurrences of life-death.[11] In his analysis of the debate between Marx and Stirner in the *German Ideology*, apropos of their common fight for life against what would be life's spectral alienation (that is, for both, the abstractions of ideology), Derrida writes:

> Both of them love life, which is always the case but never goes without saying for finite beings: they know that life does not go without death, and that death is not beyond, outside of life, unless one inscribes the beyond in the inside, in the essence of the living. They both share, apparently like you and me, an unconditional preference for the living body. But precisely because of that, they wage an endless war against whatever represents it, whatever is not the body but belongs to it, comes back to prosthesis and delegation, repetition, différance. The living ego is auto-immune, which is what they do not want to know. To protect its life, to constitute itself as unique living ego, to relate, as the same, to itself it is necessarily led to welcome the other within (so many figures of death: différance of the technical apparatus, iterability, non-uniqueness, prosthesis, synthetic image,

simulacrum, all of which begins with language, before language), it must therefore take the immune defenses apparently meant for the non-ego, the enemy, the opposite, the adversary and direct them at once *for itself and against itself*.[12]

Therefore, Marx and Stirner share the same metaphysical desire, the desire of a purely immanent life, autonomous and independent, full presence in itself, immune to what they see as external and foreign: the other, difference, the trace, etc. According to Derrida, they do not understand self-difference, the relation to the other as the necessary condition of the life of the living. Here the problems begin: the logic that regulates the life of the living in its constitution would be an autoimmunitarian logic. In order to be itself, the living must turn its autoimmunitarian defenses against itself, against the self, against the *autos* that otherwise would remain closed to the other and to difference, which the self needs in order to be itself, that is, a living being. Where does this conception of autoimmunity come from? Is there a biological foundation? Or is it just one more neologism, a mere metaphor, to say that the living must open onto the other in order to be itself, even if this opening entails mortal dangers? Let me examine the already quoted passage from *Faith and Knowledge* where Derrida gives some more precise references apropos of immunity and autoimmunity:

> It is especially in the domain of biology that the lexical resources of immunity have developed their authority. The immunitary reaction protects the "indemn-ity" of the body proper in producing antibodies against foreign antigens. As for the process of auto-immunization, which interests us particularly here, it consists for a living organism, as is well known and in short, of protecting itself against its self-protection by destroying its own immune system. As the phenomenon of these antibodies is extended to a broader zone of pathology and as one resorts increasingly to the positive virtues of immuno-depressants destined to limit the mechanisms of rejection and to facilitate the tolerance of certain organ transplants, we feel ourselves authorized to speak of a sort of a general logic of auto-immunization.[13]

Derrida tells us that he borrowed the lexicon of autoimmunity from biology. The autoimmunitary is designated as a process whose structure is quite precise: in the process of autoimmunization, the self does not turn its own immunitarian defenses against itself but protects itself against those defenses by destroying them. This means that, to a certain extent and under certain conditions, the immunitarian defenses that must protect the body from external antigens can be dangerous for the body that itself destroys them. This is quite different and above all puts us on a good track: this conception of autoimmunization would come from biology and concern the ordinary organization and behavior of the living.[14] It would find an always larger and more extensive application in medical pathology, not only in AIDS research, which would only amount to one of the possible applications, but above all in current research on immune-depressants.[15] Derrida is evidently well informed about the topic and, more generally, about the most recent areas of focus in biology. Therefore, if we cannot speak of a generic and metaphorical use, what does he mean with autoimmunity? This is my hypothesis: Derrida refers to a precise field of biology that in the 1990s had revolutionized the biological conception of life as well as the medical treatments of serious pathologies, such as AIDS and cancer, and, above all, research on immune-depressants. I refer here to the theory of apoptosis or cellular suicide. Indeed, another trace causes us to follow this direction: the recourse to the term "suicide" to describe the autoimmunitarian dynamic. In *Faith and Knowledge*, Derrida takes into account "the dimension of auto-immune and self-sacrificial supplementarity," autoimmunity as the "principle of sacrificial self-destruction ruining the principle of self-protection (of maintaining its self-integrity intact) and this in view of some sort of invisible and spectral survival."[16] In the texts that follow, Derrida explicitly recurs to suicide when referring to autoimmunity. This was evident in the passages I quoted from *Rogues*, yet it is in "Autoimmunity: Real and Symbolic Suicides" where the recourse to suicide betrays a reference to the theory of apoptosis: "An autoimmunitarian process is, as we know, that strange behavior by which, in a fashion that is quasi-suicidal, the living being 'itself' works to destroy its protections and immunize itself against itself [*à s'immuniser contre sa propre immunité*]."[17] At this point, we can turn to the theory of cellular suicide and more precisely to the contribution made in this field by the French biologist Jean-Claude Ameisen.

From Apoptosis to Autoimmunity

Ameisen first approached theories of cellular suicide when he was working on AIDS. In 1991, he published an important article titled "Cell Dysfunction and Depletion in AIDS: The Programmed Cell Death Hypothesis" in *Immunology Today*. The article was followed by a second study, published a year later in the same journal, titled "Programmed Cell Death and AIDS: From Hypothesis to Experiment." In these articles, Ameisen argues that, in case of AIDS, the destruction of the immunitarian system is not directly produced by the HIV virus, but the latter causes a deregulation in the functioning of the cellular suicide that, according to this new theory, belongs to the ordinary behavior of the organism and constitutes an essential function of the life of the living. Therefore, the organism is formed and lives thanks to a process that is not a mere cumulative growing but that implies, through different phases, from the embryo to the adult age, the death of several cellular families, a death caused by signals that come from the organism itself, and this is why we speak of suicide. The hypothesis advanced by Ameisen produced a great clamor and did not remain within the limits of specialized laboratories. In any case, there are several reasons why I believe we should follow this track: in particular, because it gives us the opportunity to interpret Derrida's autoimmunity as something more than a metaphor.

We shall read *La sculpture du vivant. Le suicide cellulaire ou la mort créatrice* (1999), the book where Ameisen proposes a unified theory of the cellular suicide that accounts for different studies carried out in multiple fields of biology and medical pathology, but which, above all, presents cellular suicide as condition of possibility of the living and of its evolution. I note that the word *deconstruction* appears in this book several times.[18]

What interests us more in this book is the interpretation that Ameisen offers of the role of cellular suicide in the constitution of the immunitarian system and of the organs with an immunitarian privilege (uterus, eyes, and brain) where the immunitarian system cannot intervene. These are evidently organs that allow for the relationship with the other in general, the environment, which is essential to the survival and the reproduction of the living.

I start from the genesis of the structure of the embryo and of the cells that compose the latter, that is, from the reproduction of life. The structure of a cell functions according to a twofold law that is apparently contradictory: reproduction and self-differentiation. In the domain of biology, the position and weight of the new cells with respect to the mother cell have long been considered as the cause of differentiation. But the development of a living organism is not a process of continuous and cumulative expansion; it implies the disappearance of an enormous number of cells and of entire cellular families:

Beyond Life and Death: Autoimmunity

The first hours, days and weeks of our life flow sheltered from the aggression of the external world, in the sanctuary of the body of our mother. However, in the absence of any disease, accident and aging, mysterious events occur: in this expanding universe, entire parts dissolve and disappear as they are built. The cells that compose the tissues and organs of the embryo are the site of massive phenomena of death.[19]

These phenomena of disappearance correspond to a sort of originary *espacement*: producing the void between the cells that is required for their displacement and the instauration of new relations between them; this hollowing out is the condition of differentiation and thus of the construction of the organism:

> A few days after fecundation, when we are only constituted by a little ball of about one hundred cells, surrounded by the superficial layer of trophoblasts that want to anchor themselves in the body of our mother, death makes brutally disappear some cells that occupy the centre and suddenly creates an empty space within the sphere. This cavity allows for the migration, the translation from periphery towards the centre, of the cells that want to transform themselves at a distance from their neighbors, and thus gives birth to new cellular families, organizes in the space the primary architecture of the body of the embryo to come.[20]

Since the 1960s, biology has begun to take into serious consideration these phenomena, designated as "cellular death," and has assigned to them an essential role in cellular differentiation and the development of the embryo, but it is only with cellular suicide that we can understand their dynamic. In fact, we have before us not a death or a disappearance due to the intervention of some actors alien to the cellular habitat, but rather the suicide of a part of the cell, or rather, of groups of cells, in response to signals that come from other cells and that are determined by the information contained in the genes, that is, caused by the programme that regulates the life of the living and not by some pathological perturbation of this programme. For this reason, scientists also speak of "programmed cellular death":[21] "It is because the cellular death is a suicide, an active phenomenon of self-destruction, and not the result of a brutal murder, a paralysis, that it can be accompanied by a discourse, by the precise emission of signals and messages, and does not occur in total silence nor in an indistinct noise, a crash."²² We came to the dynamic of cellular suicide while studying the construction of the immunitarian system and human brain, whose functions are those of constructing and protecting the identity of the living self:

> Through the study of how our two most complex and sophisticated organs are built we must detect the unsuspected power that grants to our body the control over the signals related to the programmed cellular death. In these two organs, death is at the heart of a process of learning and self-organization, whose accomplishment is neither the construction of an architecture nor the sculpture of a form but the elaboration of supports for our memory, identity and complexity.²³

The immunitarian system and the brain carry out the same function at different levels: the construction and preservation of the identity of the living, that is, the construction of the physiological and nervous memory that presides over the relationship of the living with its internal and external environment:

> At first glance, these two organs do not have much in common. However, at different degrees and in a very different way, the brain and the immunitarian system share a property that is essential and mysterious at once: to ensure the permanence of our singular identity, to adapt us to the environment and construct in us a memory that will convert the innumerable sequence of aleatory and contingent events that have happened to us into a history, our history, and will allow us to decipher the present and project us into the future in the light of a past that is recomposed without end. These notions of identity, adaptation and memory imply the capacity of perceiving the modifications of our internal and external environment as information, of distinguishing the one from the other, of integrating, responding and being able to recognize them in the future. . . . The notion of memory presupposes the capacity of distinguishing between new information we have never encountered before, and information that we have already encountered. Remembering is recognizing; and recognizing is responding to the second time in a way that will be different from the first one. But these two notions of identity and memory overlap and integrate one another. We

modify ourselves perpetually as much as the environment in which we are immersed modifies itself.[24]

There is no need, I believe, to search for a deeper congruency between this conception of memory as the genetic and structural condition of the individuality of the living and Derrida's notion of arche-writing as it has been described and employed throughout this work. For the moment, I prefer to focus on the role played by cellular suicide in the immunitarian system. The structure and the functioning of our immunitarian system depend on the interaction between two different cellular families, lymphocytes and guardian cells: "It is the family of the T lymphocytes, produced by the thymus as the embryo develops, that plays an essential role in the recognition of the non-self and in triggering and coordinating the fights against the microbes."[25] There are hundreds of millions of T lymphocytes in us, and they are all different from one another. Each lymphocyte carries on its surface thousands of samples of a special structure that allows it to explore the environment—the receptors through which it can perceive, identify, and respond to the microbes that invade us. But the T lymphocytes do not react directly to the microbes that come from outside. They first need to recognize the difference between the self they must protect and the microbes that do not belong to the self. This recognition hinges on the interaction between lymphocytes and guardian cells. The latter carry on their surface certain displays on which a large variety of proteins produced by the cells of the organism are fixed:

> In absence of infection, the only proteins that the guardian cells contain and whose fragments they expose on the surface of their displays are the proteins produced by our body on the basis of the information contained in the genes. It is the sum of the assemblages of our displays and a fragment of each one of our proteins that, for our immunitarian system, constitutes the self, our identity. But when a microbe invades our body, these guardian cells that the microbe penetrates or by which it is captured start to cut out a part of the proteins that compose them and to expose, on their surface, on the displays, fragments of these proteins stranger to our body. . . . This assemblage is made of a portion of non-self (a fragment of protein coming from a virus, a bacterium, a parasite) inserted in a portion of self (the display). In this way the intrusion, the presence of

the aggressor, is revealed, deciphered, identified in the context of the self. The capture of this combination of non-self and self, by the receptor of a lymphocyte T, is indispensable to its transformation in the fight.[26]

Cellular suicide has been onstage since the first days of the life of the embryo and thus of the production of the immunitarian system, precisely when the selection of the lymphocytes able to dialogue with the guardians takes place. There may be some lymphocytes that are unable to communicate with the guardians or to recognize the self and thus are useless, but, above all, there are other lymphocytes that recognize the self, and only the self, too well, so as to identify the self's presence alone and ignore the non-self being displayed. Once active in the life of an accomplished organism, these lymphocytes will become dangerous for the self they must protect: unable to recognize the difference between the self and the non-self, they can confuse the one with the other. These lymphocytes are too attached to identity—like Marx and Stirner, we may say—and do not recognize the traces of difference. Therefore, if they survive in the accomplished organism, they provoke autoimmune diseases. To sum up, the organism must destroy two types of lymphocytes. In particular, it must destroy the immunitarian defenses that are too powerful, as they are dangerous for the life of the living, and it does this through the signals of cellular suicide that are sent to the guardians. Here we may recall what Derrida said: "An autoimmunitarian process is, as we know, that strange behavior by which, in a fashion that is quasi-suicidal, the living being itself works to destroy its protections and immunize itself against own immunity [*à s'immuniser contre sa propre immunité*]."

But we go further with our reading of Ameisen:

> Any lymphocyte whose receptor interacts too well with one of the assemblages that constitute the self will risk to attack one day the body it belongs to, destroying a tissue or an organ: it is a lymphocyte that shows its potentially dangerous nature and would be able to provoke a disease called "autoimmune." Conversely, any lymphocyte whose receptor is totally unable to interact with any assemblage that constitutes the self (and thus with any display of the guardian cells) will never be able to interact, after the birth of the child, with the display of a guardian cell, on which fragments of microbes are exposed. It

is a lymphocyte that shows its probable inability to protect the body.... In the body of the embryo in construction, the individual destiny of each lymphocyte T, its survival or death, will depend on the nature of the interactions between its receptor and its environment. Any excessive attachment of the receptor to the self exposed by the guardian cells triggers a strong signal that provokes the immediate suicide of the lymphocyte that receives it. The lymphocyte disappears at the precise moment it proves its dangerous character. Conversely, a receptor that cannot interact with the self at all will be unable to transmit over three days a signal to the lymphocyte by which it is carried. The absence of signal will trigger the suicide of the lymphocyte that proved its inability to interact with the guardian cells.[27]

Above all, we must understand the role of cellular suicide in the constitution and the defense of the organs with immunitarian privilege, the three organs essential to the life of the living, in which the immunitarian system cannot intervene. These are the uterus, the eyes, and the brain. They are essential as they are open to the alterity, the non-self, and necessary to the constitution of the identity of the living, to survival and reproduction. Until the focus on cellular suicide in the 1990s, we previously believed that the organism defends this privilege against its own immunitarian system. Until then we believed that there were only physical barriers that prevented lymphocytes from penetrating; now we know that cellular suicide is also at work here:

This privilege does not only result from the existence of a passive barrier that prevents the immunitarian system from accessing to a territory. It also results from the existence of an active barrier, constituted by cells that trigger the suicide of the fighters of the immunitarian system and thus brutally interrupt their afflux and prevent them from advancing in the sanctuary where the microbe that they are tracking throughout the body escaped. Even if this sounds strange, our body and immunitarian system are engaged in a mortal combat. The songs of death, the signals of the cellular suicide, are the veritable guardians of the sanctuary.[28]

I believe that we should not overlook the traits shared by Ameisen's and Derrida's discourses on the defense of the living in relation to the autoimmunitarian system. However, this does not mean to blur distinctions, or to make Derrida's discourse derive from Ameisen's. The latter interests us so long as it permits us to grasp the scientific consistency of Derrida's discourse while safeguarding the originality of its bearing.

In the concluding paragraphs of the chapter that Ameisen devotes to the role played by cellular suicide in the constitution of the self and the immunitarian system, the analogies are even more suggestive. Through them, we may reconsider Derrida's recourse to the autoimmunitarian lexicon from another perspective:

> The cells of the sanctuaries force to the suicide those fighters that rush there to hunt down the potentially dangerous dissonances they perceived on the melodic line of the self. The coexistence of our body and our immunitarian system results from a dynamic balance, a permanent fight. A body that protects itself too well against the guardian by destroying it will destroy itself. An immunitarian system that will protect too well the body by destroying it will destroy itself. The maintenance of our integrity and identity results from a complex relation of forces, from an armed peace interrupted by sudden fights.[29]

Here we have a discord that is fought by means of iterable traces, signals that induce suicide in the enemy who in turn responds with other signals that could neutralize them by suspending and temporarily deferring death. For this reason, Ameisen concludes: "A strange vision of life is at stake here: living, for each cell . . . is having persistently succeeded, for some time, to repress the activation of its suicide."[30]

> Life does not let itself be separated from death. It is inhabited by death. The cells of the embryo . . . are ready to destroy themselves from the beginning to end of their period of development. The universe of the embryo that constitutes itself is at any moment ready to activate its premature collapse. It bears within itself its negation, disappearance, effacement, like a potentiality that is always open. Since its birth, each cell has been a few instants far from its possible death. Only the fabrication of the protector allowed it to continue the journey. Survival for an embryonal cell means repressing the death in itself, that is, at each instant, prolonging a delay, giving itself a break [*prolonger un sursis, s'accorder un répit*].[31]

Therefore, life "consists of the conquest, day by day, of an ephemeral suspension that must be ceaselessly renewed."[32] The living must protect itself from a too-rigid immunitarian system, that is, from a too-strong identity, that would be closed in on itself, autistic, deaf to the signals of difference that come from alterity, that is, from death, which it bears within itself. At the same time, the living must not completely renounce its defenses in relation to the other as long as its own identity and thus its life are at stake. The opening onto the other is necessary to life but, as it also entails death, this opening opens at the heart of the living a space that is not pacific but differential, relational, conflictual, the space of the constitution of our identity where life gains its survival in order to delay, postpone, and defer the end of the play between the body and the immunitarian system, between life and death, identity and difference, as long as possible. In other words, to avoid one of the two terms winning and annulling (the relation to) the other: a perfectly close and absolutely self-adherent identity and the absolute absence of identity are exactly the same: death. From this perspective, a full life, purely immanent in itself, intact and immune to the other and to difference, the life desired by Marx and Stirner, heirs of the long metaphysical tradition, would not merely be a philosophical error but the constitution of a true threat for life itself, the most dangerous threat. Therefore, we can finally respond to the question raised at the beginning: why does Derrida speak of the autoimmunitarian apropos those identitarian processes that are not the processes of the living but of consciousness, reason, community, and democracy? To conjure away the mortal danger that threatens these forms of life from within: the metaphysical desire or phantasm of an absolute and unconditioned identity, that would be healthy, safe, intact, pure, immune.[33] To conjure away the pathological autoimmunity that affects our cultural body (symbolical, religious, political, and democratic) by forcing it to immunize itself absolutely from the other, the other that it needs in order to be what it is, a form of life, at least according to the logic of the processes of autoimmunization that condition the organization and the survival of the living in its ordinary biological development. In the passage that follows, Derrida draws attention to the example, if not the law, of that development:

> If an event worthy of this name is to arrive or happen, it must, beyond all mastery, affect a passivity. It must touch an exposed vulnerability, one without absolute immunity, without indemnity; it must touch this vulnerability in its finitude and in a nonhorizontal fashion, there where it is not yet or is already

no longer possible to face or face up to the unforeseeability of the other. In this regard, autoimmunity is not an absolute ill or evil. It enables an exposure to the other, to *what* and to *who* comes—which means that it must remain incalculable. Without autoimmunity, with absolute immunity, nothing would ever happen or arrive; we would no longer wait, await, or expect, no longer expect one another, or expect any event.[34]

It is from this point, from the "strange illogical logic"[35] of autoimmunity, that we must reconsider the question of the community in view of the possibility of its survival:

> Community as *com-mon auto-immunity*. No community is possible that would not cultivate its own auto-immunity, a principle of sacrificial self-destruction ruining the principle of self-protection (of maintaining its self-integrity intact) and this in view of some sort of invisible and spectral survival. This self-contesting attestation keeps the auto-immune community alive, which is to say, open to something other and more than itself: the other, the future, death, freedom, the coming or the love of the other, the space and time of a spectralizing messianicity beyond all messianism.[36]

VII

Living On: The Arche-performative

"Always prefer life and constantly affirm survival . . . I love you and am smiling at you from wherever I am."[1]

These are the concluding lines of the short handwritten text that was read on October 12, 2004, on the occasion of Jacques Derrida's funeral, and distributed to the attendants. The reading of this text cannot end with the moment of mourning. The question of survival is eminently theoretical, it consists in the heart of deconstruction. It beats secretly under deconstruction, as it develops. By survival, Derrida understands the movement of *différance* as the irreducible and structural condition of the life of the living, behind the supposed opposition of life and death, which has always organized and oriented the determination of the meaning of these terms. In order to demonstrate this hypothesis, we return to Derrida's interpretation of linguistic acts, following a determinate path that is worth reconstructing step by step: (1) this interpretation searches for the condition of possibility of the performative linguistic act; (2) it focuses on a certain experience of the performative dimension of writing (in particular and for essential reasons, of literature) in which (3) it is possible to look at survival as the irreducible structure of the living. An irreducibly performative structure, according to a dimension of performativity that is different from and yet related to the performativity of linguistic acts insofar as it consists in the latter's irreducible condition of possibility. Before proceeding along this track, so as to breach a way—a *via rupta*—I shall establish some more general coordinates. Derrida explicitly affirms the theoretical relevance of the notion of survival for his work. I shall read the last words of *Learning* to Live Finally, the interview given by Derrida to Jean Birnbaum when death was imminent:

> As I already said, from the outset and long before the experience of survival that are actually my own, I noted that survival is an original concept, which defines the structure itself of what we call existence, the Da-sein, if you will. We are, structurally speaking, survivors, marked by this structure of the trace, of the testament. That said, I would not endorse the view according to which survival is defined more by death, the past, than by life and the future. No: deconstruction is always on the side of the affirmative, the affirmation of life. Everything I have said at least since Steps (Parages, 1986) about survival as a complication of the opposition death-life proceeds with me from an unconditional affirmation of life. Survival is life beyond life, life more than life, and the discourse I undertake is not death-oriented, just the opposite, it is the affirmation of someone living who prefers living, and therefore survival, to death; because survival is not simply what remains, it is the most intense life possible.[2]

Survival is the irreducible, structural condition of existence. I note the generality of the term "existence" and the possible alternative: the Heideggerian notion of *Dasein*. Like *Dasein* and, on behalf of it, survival precedes and conditions the determination of human individuality as psyche, soul, subject, speculative or phenomenological consciousness, and so forth. In order to understand survival as the irreducible and structural condition of existence, it is necessary to detect, on the one hand, its matrix (the differential trace) and, on the other, its affirmative bearing (the *yes*, the unconditional affirmation of life). As I shall point out, in order to understand survival and, thus, to affirm it, as Derrida himself conjures us to do, it is necessary to get through a certain experience and interpretation of the performative. "Always prefer life and constantly affirm survival . . ." is a performative statement: it does not describe anything but is an order, a prayer, an appeal to swearing or promising, it is somehow brought out in the instant of death. Beyond this evidence, which is, perhaps, trivial, the notion of survival as the structural condition of life rests on a certain interpretation of the condition of possibility of the linguistic performative. In particular, I follow the traces of the *yes*. In "A Number of Yes," published in 1987, Derrida understands the *yes* as the exemplary performative by means of which we may go back to a "quasi-transcendental or ontological"[3] performativity that, within any linguistic act, within the constative-performative opposition, would be the condition of possibility of all utterances. However, in order to find in this quasi-transcendental performativity the structure of the survival, it is necessary to turn to literature. Once again in an interview—and this is the second coordinate of our path—Derrida explicitly affirms the link between his concern for literature and the instance of survival:

> What counted for me [in literature] is the act of writing or rather since it is perhaps not altogether an act, the experience of writing: to leave a trace that dispenses with, that is destined to dispense with the present of its originary inscription, of its "author" as one might say in an insufficient way. This gives one a way that is better than ever for thinking the present and the origin, death, life or survival. Given that a trace is never present without dividing itself by referring to another present, then what does being-present, or the presence of the present mean? The possibility no doubt carries beyond what is called art or literature, beyond in any case the identifiable institutions of that name.[4]

Literature is the major way to grasp the articulation between the structural condition of life and that of writing, in which life is inscribed.[5] Therefore we shall trace the experience of writing as it unfolds itself in literature in order to account for the structure of survival, with a focus on literature that questions the sense of the experience of writing. The reference to *Parages* in the above-quoted passage from *Learning to Live Finally* leads us to look at Maurice Blanchot in our search for the literature that polarizes Derrida's interest. The central part of *Parages*, which is dedicated to Blanchot, is the essay "Living On," first published in English in a well-known collective volume.[6]

The Double Yes:
From the Linguistic to the Arche-performative

In "A Number of Yes," Derrida highlights in Michel de Certeau's work a certain engagement with the *yes* that takes the latter from the performative linguistic act of the stated "yes" to a more original performative, another *yes*, which is presupposed by all linguistic acts, without being ever uttered, and stands for their condition of possibility:

> What he has said to us *on the subject* of the *yes* was not simply a discourse on a particular element of language, a theoretical metalanguage bearing on one possibility of utterance, on one scene of utterance among others. . . . Because a *yes* no longer suffers any metalanguage; it engages the "performative" of an originary affirmation and remains thus presupposed by every utterance *on the subject* of the *yes*. Moreover—to put it aphoristically—for Michel de Certeau, there is no subject of any kind that does not arise from the scene of the *yes*. The two *yesses* we have just discerned (but why are there always two? We will ask ourselves this question again) are not homogeneous, and yet they are deceptively similar. That a *yes* should be presupposed each time, not only by every statement on the subject of the *yes*, but also by every negation and every opposition, dialectical or not, between the *yes* and the *no*, this is perhaps what immediately gives the affirmation its essential, irreducible *infinity*.[7]

Therefore, we may isolate the conditions of possibility of any linguistic act only by going from the linguistic performative of the *yes* to the original and non-linguistic one, to the original affirmation that consists in the condition of possibility of existence and thus of life. In fact, Derrida explains that for de Certeau, this original affirmation anticipates and conditions the very possibility of the constitution of the subject (psyche, soul, consciousness . . .). On the other hand, Derrida himself retraces the original affirmation back to the transcendental dimension when he determines as "arche-originary" the *yes* "that gives the first breath to every utterance [*qui donne son premier souffle à toute énonciation*]."[8] In so doing, he posits the original affirmation at the level of arche-writing (*archi-écriture*) and arche-trace (*archi-trace*), namely, of the structural or "quasi-transcendental"—but not *a priori*—conditions of possibility. That is why this notion of original affirmation resists the Heideggerian objection according to which it would be the ultimate development of the "metaphysics of the will," the modern declination of the "metaphysics of the presence."[9] The original affirmation anticipates by far the constitution of any form of subjectivity, that is, of the necessary support to the exercise of will. It precedes that constitution as its very condition, remaining itself "unconditioned"[10] and, thus, independent from any voluntary, conscious, and self-present deliberation:

The archi-originary *yes resembles* an absolute performative. It does not describe or state anything but engages one in a kind of archi-engagement, alliance, consent, or promise that merges with the acquiescence given to the utterance it always accompanies, albeit silently, and even if this utterance is radically negative. Given that this performative is presupposed as the condition of any determinable performative, it is not simply one performative among others. One could even say that, as a quasi-transcendental and silent performative, it is removed from any science of utterance, just as it is from any speech act theory. It is not, strictly speaking, an act; it is not assignable to any subject or to any object. If it opens the eventness of every event, it is not itself an event. It is never *present* as such. What translates this nonpresence into a present *yes* in the act of an utterance or in any act at the same time dissimulates the archi-originary *yes* by revealing it.[11]

In "A Number of Yes," Derrida sketches out the general traits of this "arche-originary affirmation,"—engagement, alliance, promise—without describing its dynamic and structure. Only indirectly may we account for its articulation with the structural condition of existence and, thus, of life. From this perspective I shall refer to literature, or, rather, to narrative (*récit*), to a singular narrative: Blanchot's *Death Sentence* (*L'arrêt de mort*, 1948).

Living On: The Arche-performative

Before taking this step forward, let me clarify the exemplarity of narrative (*récit*) in general, of the narrative's origin and structure. Derrida insists on the term "*récit*," which in English is deprived of the original, semantic density. The French "*récit*" refers to "repetition," "quotation," "recitation," to the performance of an attestation in which something passed away is repeated. The origin and the structure of narrative (*récit*) imply the call for a testimonial act with respect to a past present. A narrative is structured around such a call, which, somehow, consists in its very origin and condition:

> I suggest, for example, that we replace what might be called *the question of narrative* [récit] ("What is a narrative?") with *the demand of narrative*. When I say *demand* I mean something closer

to the English "demand" than to a mere request: inquisitorial insistence, an order, a petition. To know (before we know) what narrative is, the narrativity of narrative, we should first perhaps recount, return to the scene of one origin of the narrative (will that still be a narrative?), to that scene that mobilizes various forces, or if you prefer various agencies or "subjects," some of which *demand* the narrative of the other, seek to extort it from him like a secret-less secret, something that they call the truth about what has taken place: "Tell us exactly what happened." The narrative must have begun with this demand, but will we still call the *mise en scène* of this demand a narrative?[12]

Therefore, a narrative has to do with the present attestation of a present which is already past at the moment of the attestation. From a phenomenological perspective, I can go back to the living presence of the past present through the ideal objectivities—traces, signs, words—that refer to it. According to Derrida, this is the metaphysical illusion of phenomenology, as well as of any philosophy founded on the immediacy of the presence, including speech act theory. Conversely, only the possibility of a trace that is absolutely different and independent from the immediate presence it refers to as already passed away, only this possibility grants the constitution of the present, even if always just in view of a further reference, a reference to come or a deferred one. Appealing to phenomenological language, I shall say that, since the beginning, for the consciousness that is constituted through the retention of a trace of experience, there has been no trace of the living present of experience but only the trace of the deferring of (from) this present, a trace oriented toward the to-come.[13]

Therefore, Derrida is interested in narrative in general and, in particular, in Blanchot's novels. In the aforementioned passage from "Living On," he refers to *The Folly of the Day* (*La folie du jour*, 1973) in order to recall the interpretation proposed in "Pas," the first essay included in *Parages*. In "Pas," Derrida draws attention to the staging of the demand of narrative as the very origin of narrative itself and thus of the latter's institutionalization in what we call literature. In *The Folly of the Day*, the narrative "I" must respond to what happened, must tell the truth about what was present, and yet affirms that it does not remember at all. The injunction comes from a doctor and a police officer, institutional instances that ask the narrative voice to institute itself as a narrative subject within the order of the metaphysics of presence. Only the possibility of returning to the living presence of the past present grants the truth of the ideal objectivities that refer to it. A witness must be able to re-present the living present in which he/she was present. In staging the law imposed on the novel from outside, *The Folly of the Day* solves the narrative itself from this bond and shows, through the very forward march of narrative, the impossibility of such a presence and, at the same time, the necessity of a more in-depth search for the ultimate condition of attestation, which touches on the very heart of life. For this reason, instead of dealing with Shelley's *The Triumph of life*, Derrida reads *Death Sentence*:

> Within the boundaries of this session, I shall propose a fragment, itself unfinished, detached from a more systematic reading of Shelley, a reading oriented by the problems of narrative [récit] as *reaffirmation* (yes, yes) of life, in which the *yes*, which says nothing, describes nothing but itself, the performance of its own event of affirmation, repeats itself, *quotes*, *cites* itself, says *yes-to* itself as (to an-) other in accordance with the ring [anneau], requotes and recites a commitment [engagement] that would not take place outside this repetition of a performance without presence. This strange ring says *yes* to life only in overdetermining ambiguity of the triumph "of," "over" life, "over," "on" [sur] life, the triumph marked in the "on" of "living on" [dans le *sur* d'un survivre].[14]

Once it is liberated from subordination to the institution called literature and, thus, to the order of the metaphysics of presence, *The Folly of the Day* reveals in an exemplary fashion the structure of the double *yes* as the structural condition of attestation. Given the absolute alterity of the trace with respect to the living present to which it refers, narrative attests to the necessity of the self's reference (*renvoi*) to the other as other, to what is other than the living present—the iterable trace—in order to relate to itself and thus to be itself. From this perspective, the double *yes* consists in the self-affirmation that must go through the repetition of the trace, through the confirmation of the trace as the trace of the self. This self-affirmation through the confirmation of the traces elaborated by experience is necessarily implied in all linguistic acts and constitutes their condition of possibility. For Derrida it is necessary to demonstrate that, through the reading of *Death Sentence*, (1) the structure of the double *yes*, which is presupposed by all speech acts, is an articulation and an effect of the irreducible structure of the life of the living; (2) that the latter depends on the arche-originary *yes*, an unconditional *yes* to life that structures the life of the living before its constitution as a subject, as a consciousness that reappropriates itself to itself in its presence in a punctual and living present; (3) that, on the basis of this unconditional *yes* to life, the dynamic of the life of the living must be described in terms of survival. Finally, it must be possible to find in survival itself the arche-performative.

Derrida insists on the semantic density of the title, which is lost in the English translation *Death Sentence*. The French *"arrêt de mort"* certainly means "death sentence," but *"arrêt"* alone also means interruption, suspension, pause, arrest. Furthermore, Derrida observes that the occurrence of *"arrête"* already comprehends *"arête,"* with one "r," which refers to a line of contact or intersection:

> Arrête, with two "r," is thus indeed that which orders the *arrêt* (stopping/decision), but the *ar(r)ête*, as a noun, is also that sharp dividing line [limite aiguisée], that angle of instability on which it is impossible to settle to *s'arrêter*. Thus this dividing line functions also *within* the word and traces in it a line of vacillation. This line runs within *L'arrêt de mort*, within what the *arrêt de mort* says, the expression "arrêt de mort," the title *L'arrêt de mort*—all of which are to be distinguished.[15]

We are not before a mere linguistic speculation. According to Derrida, narrative itself asks us to reckon with the semantic density of its title: it is divided into two parts that are linked by no evident diegetic relation. The first part develops into two main episodes: a first one, in which (the sentence of) death is suspended, deferred, life goes on, and another one, in which (the sentence of) death is definitive, or, rather, so it seems, because in fact the reader remains unsure about death: is it definitive or still deferred? It is clear that the two episodes send the one back to the other so as to render the line that should separate them (as well as life and death) unstable and permeable. Both episodes are referred to by the narrator, who speaks in the first person (*Je*); they concern the same person, a woman, a friend, perhaps a lover of the narrator, who calls her by what may be the initial of her name: J. (in a relation of homophony with *Je*). From the beginning we know that the woman is affected by an incurable disease and "should have been dead already. She thus lives *on* [*Elle sur-vit donc*]."[16] She survives her own death till she makes the decision (*l'arrêt*) to die with the help of the narrator, but only later, after the episode of the delay of a death that seemed to be unavoidable, will the narrator decide to give death to her, even if he is no longer sure of the woman's consent. Therefore, death remains uncertain, and the novel stops (*s'arrête*) by leaving it definitively suspended. Death is never present even if remains *present* in any instant of J.'s life. In this suspension (*arrêt*) that interrupts and delays (*arrête*), that defers the already decided death (*l'arrêt de mort*), Derrida finds the dynamic of différance at the heart of life: the mutual, differential relation between life and death that structures and orients the movement of the narrative and by means of which it is possible to understand the dynamic of the life of the living as survival:

> The *arrêt de mort* is not only the decision that determines what cannot be decided: it also arrests death by suspending it, interrupting it, deferring it with a "start," the startling starting over, and starting on, of living on [la diffère dans le sursaut d'un survie]. But then what suspends or holds back death is the very thing that gives it all its power of undecidability—another false name, rather than a pseudonym, for différance. And this is the pulse of the "word" *arrêt*, the arrhythmic pulsation of its syntax in the expression *arrêt de mort*. *Arrêter*, in the sense of suspending, is suspending the *arrêt*, in the sense of decision. *Arrêter*, in the sense of deciding, arrests the *arrêt*, in the sense of suspension. They are ahead of or lag behind one another. One marks delay; the other, haste. There are not merely two senses or two syntaxes of *arrêt*, but beyond a playful variability, the *antagony* from one *arrêt* to the other. The antagony lasts from one to the other, one relieving the other in an *Aufhebung* that never lets up, *arrêt* arresting *arrêt*, both senses, both ways. The *arrêt* arrests itself. The indecision of the *arrêt* intervenes not *between* two senses of the word *arrêt* but *within* each sense, so to speak.[17]

I follow Derrida's reading of the episode where J. seems to be dead and yet comes back while the narrator (*Je*), who is certain of the ineluctable, calls her by her name. This allows us to detect the dynamic of différance at the heart of life and, finally, to touch on the unconditional *yes* to life as the irreducible condition of this dynamic. As she survives death, J. cannot remember what happened, even if her state has changed; suddenly she looks happy with the delay of her own death, which, however, she ignores:

> The reaffirmation, the *récit* of life marks its discreet triumph in a "gaiety" (the words "gay" and "gaiety" recur five or six times) the memory of which is terrifying, would "be enough to kill a man." Gaiety, reaffirmation, triumph *over* (triumph of the "on," "over," *sur, hyper* . . .): over life and of life, life after life [*sur-vie*], at the same time between life and death in the crypt, more than life [plus-que-vie, plus-de-vie], when it's over (*and over again*), reprieve [*sursis*] and hypervitality, a supplement of life that is *better* than life *and better* than death, a triumph of life and of death; a living-on [survie] that is better than truth and that would be *la chose par excellence: sur-verité*, truth beyond truth, truth beyond life and death.[18]

The unconditional *yes* to life, the reaffirmation of life is not the manifestation of the will to life, which would presuppose a self-present consciousness able to affirm its own will. In fact, J. does not remember anything: J. has never been present to what has happened to her (that is, conscious), and, yet, without being aware of it, she is happy of having delayed a death that has never been present as such. According to Derrida, this happiness without consciousness is the very manifestation of the unconditional *yes* to life, of the arche-performative that structures the life of the living at a biological, natural, and thus unconscious level, before and independently of any constituted subjectivity or consciousness:

> There is a great deal to be said about this gaiety, about the quality of the experience thus designated to describe what is proper to an act or instance of living on, the levity of its affirmation, of the *yes, yes*, of the *yes to yes* [du *oui, oui,* du *oui à oui*] without self-recollection [sans mémoire de soi], the *yes* that, saying and describing nothing, performing only this affirmation of the *yes* saying *yes to yes, must not even* have, and know, itself [s'avoir et se savoir]. But this "need not" or "must not" is also an interdiction that interposes an unconscious between the event and the very experience of it, between the living-on and the present, conscious, knowing experience of what that comes about [de ce

qui arrive ainsi]. I—the one who says me, that is to say, me—do not know what has happened, what will have happened to me.[19]

Through the story of J. and Je we learn that the unconditional *yes* to life says *yes* to the injunction to live that comes from the to-come and structures the life of the living by pushing it to live on beyond the present: life, or arche-performativity. The logic immanent in the living is not the conservation of identity, of self-presence, which would be pure and immune from alterity or difference: it is not the conservation of life before death as a so-called natural accident, external and contingent with respect to life:

> This "vivre, survivre" delays at once life and death, on a line (the line of the least sure *sur-*) that is thus one neither of clear-cut opposition nor of stable equivalence. "Living, living-on" differs and defers, like différance, beyond identity and difference. Its domain is indeed in a narrative [récit] formed out of traces, writing, distance [éloignement], teleo-graphy. Tele-phone and tele-gram are only two modes of this teleo-graphy in which the trace, the grapheme in general, does not come to attach secondarily to the telic structure but rather marks it *a priori*.[20]

Once it is conceived of on the basis of survival, the logic of the living seems to have a telic structure; it is its own self-reaffirmation, that is, the differential iteration of itself beyond the present toward the future.

Traces of Life in the Text

On the account of the notion of survival, it is time to reflect on the articulation between the unconditional *yes* to life that structures the life of the living, on a natural and biological level, as reproduction and self-repetition, and the ideal objectivities (traces, signs, and words) that consciousness elaborates in order to relate itself to experience and to give a meaning to life. The narrator (Je), who, like a witness, should refer to the event that has occurred, attests to his being present at the occurrence of the past event and, thus, finds himself in the same condition as J. He is a survivor: in his narrative there is no trace—and there couldn't be—of an event in its present and punctual occurrence; there is only a trace of iterable traces—signs, words—elaborated in order to refer to an event in a posterior moment, the moment of narration, which is absolutely other than the present to which it cannot relate and yet, exactly for this reason, can ensure the survival to come:

> The unnarratable event of J.'s coming back to life [de la survie] holds the *récit* breathless for an interminable lapse of time that is not merely the time of what is narrated: the one who narrates is also, first, *one who lives on* [survivant]. This living on is also phantom revenance (the one who lives on is always a ghost) that is noticeable (re-markable) and is represented from the beginning, from the moment that the posthumous, testamentary, scriptural character of the narrative [récit] comes to unfold.[21]

The absolute inaccessibility of the living present, necessary to the elaboration of an iterable trace, does not affect the possibility of testimonial attestation. Conversely, it makes attestation possible, it demands for attestation according to the telic and performative structure of survival: "Différance—*arrêt de mort* or triumph of life—defers (differs like) the narrative of (from) writing [*La différance, arrêt de mort ou triomphe de la vie, diffère (comme) le récit d'écriture*]."[22] By means of survival it is possible to rewrite the genesis and structure of the elaboration of the ideal objectivities—traces, signs, words—according to the opening of a to-come in the name of which we should engage our testimonial attestation, according to the protention that structures subjectivity through its irreducible, biological determination. In particular, for Derrida, the structure of survival allows us also to reconsider the questions of writing and reading, of the translation and interpretation of texts and, thus, of the transmission of the legacy they represent for us:

> A text lives only if lives *on*, and it lives *on* only if it is *at once* translatable and untranslatable (always *at once, and: ama*, "at the same time"). Totally translatable, it disappears as a text, as writing, as a body of language. Totally untranslatable, even within what is believed to be one language, it dies immediately. The triumphant translation is neither the life nor the death of the text, only or already its living *on*, its life after life. The same thing will be said of what I call writing, mark, trace, and so on. It [ça] neither lives nor dies; it [ça] lives *on*. And it [ça] "strats" only with living *on* (testament, iterability, remaining [restance],

crypt, detachement that lifts the strictures of the "living" *rectio* or direction of an "author" not drowned at the edge of his text).[23]

Derrida will return to these issues in the 1990s, for instance, in *Specters of Marx*, in view of accounting for the irreducibly spectral dimension of ideal objectivities; in *Faith and Knowledge*, "Poetics and Politics of Witnessing," and "Demeure," where testimonial performativity is understood as the condition of attestation. And it is in "Demeure," which focuses precisely on Blanchot's "The Instant of My Death" (*L'instant de ma mort*), that the question of survival comes back in relation to testimony.[24]

Living Ouverture

To conclude, we may say that the logic of survival, which is immanent in the biological structure of the living, consists in the self-protection through the traces of alterity in which the living binds itself to itself and affirms itself as such, and thus as a spatio-temporal differing (*espacement*) through traces that, being produced in view of a reference to come, must differ from the living present. To support this thesis, which might sound hasty, I refer to a few key moments in Derrida's work that anticipate "Living On." In "Freud and the Scene of Writing," published in 1966, Derrida already acknowledged the necessity of the iterable structure as the answer to the logic of survival:

> No doubt life protects itself by repetition, trace, différance (deferral). But we must be wary of this formulation: there is no life present at first which would then come to protect, postpone, or reserve itself in différance. The latter constitutes the essence of life. Or rather: as différance is not an essence, as it is not anything, it is not life, if Being is determined as *ousia*, presence, essence/existence, substance or subject. Life must be thought of as trace before Being may be determined as presence.[25]

It is not by accident that on this occasion Derrida identifies survival with the function of the iterable trace, with the inscription retained in the memory that precedes any recourse to empirical writing (*arche-writing*): "We must account for writing as a trace which survives the scratch's present, punctuality, and stigme. . . . Writing supplements perception before perception even appears to itself (is conscious of itself). 'Memory' or writing is the opening of that process of appearance itself. The 'perceived' may be read only in the past, beneath perception and after it."[26]

From this perspective, it may be even more interesting to reread Derrida's unedited seminar *La vie la mort* (1975). In the first part, he addresses the notions of "trace," "code," "program," and "text" in contemporary biology. In particular, he refers to François Jacob's *The Logic of Life* (1970). As Derrida remarks, Jacob describes genetic heritage as "a sequence of chemical radicals":[27] "Heredity becomes the transfer of a message repeated from one generation to the next. The programmes of the structures to be produced are recorded in the nucleus of the egg."[28] The logic of the living is a logic of retention and reproduction, archive and transmission. It is exactly from this perspective that we may think of survival as the performative structure of the living: "All begins with self-reproduction,"[29] Derrida states categorically. He is even more precise: Jacob's description of the logic of the living imposes the logic of différance, trace, and text as an irreducible necessity. It is not a simple metaphor or a vague analogy:

> I say *text* and not speech, not an a-textual verbal language. It goes without saying that the genetic text is not verbal, that it is aphonic, that is not what I want to stress here. . . . And that is why the notion of text imposes itself on life-science, not only imposes itself more than the notion of verbal language—that goes without saying since there is no voice and no words in genetic programmes, but—which goes less without saying for biologists such as Jacob and others—imposes itself even more than the notion of message, information and communication. There are of course effects of message, information and communication, but on condition that they be in the final analysis textual, i.e. that the message, the communication, the information never transmit (never emit, communicate, inform) any content that not itself be of the order of message, information, communication, thus not be itself a trace or a gramme.[30]

At this point, once we have detected the earliest traces of survival in Derrida's work, we must go back to the beginning, to what at the beginning could appear as the end and, thus, relaunch the opening onto the to-come: "Always prefer life and constantly affirm survival. I love you and am smiling at you from wherever I am." Reaffirming survival means to affirm the unconditional *yes* to life, before the traditional opposition between life and death, before a conception of life as a full presence or the punctual self-identity of a supposed living, absolute and unconditioned present. Such a life could only be attributed to the God of metaphysics, in its historical variations, and, once projected or imposed on the natural living being, it would merge with death itself, with its denegation. Survival consists in the irreducible condition of possibility of the life of the living but also, at the same time, in what irreducibly exposes life to death. Eluding the possibility of death would also mean removing the possibility of life. It is therefore in the name of life, of another conception and experience of life, that Derrida binds our attestation to a survival to come. The possibility of the to-come, in contrast with the supposed living present and its institution, is the dimension that structures the life of the living. It is to the structure of this performative *ante litteram*, which I designated as "telic," that we should refer in order to rethink life as well as the structures and institutions in which it is inscribed: the genesis and structure of what we conventionally call subjectivity as well as the institutions of knowledge, politics, economy, religion, society, and so forth, which are established on a notion of subjectivity that, at this point, seems to be unable to account for them.

Notes

Introduction

1. Jacques Derrida, *Learning to Live Finally: The Last Interview*, 51.
2. Derrida, *For What Tomorrow*, 63.
3. Derrida, "From Restricted to General Economy," 441.

Chapter I

1. On the notion of "arche-writing," permit me to refer to Vitale's entry on *"archi-scrittura* [arche-writing]" in Facioni et al., *Derridario. Dizionario della decostruzione*, 15–28.
2. Derrida, *Edmund Husserl's Origin of Geometry: An Introduction*, 85.
3. Derrida, *Speech and Phenomena*, 85.
4. Derrida, *Edmund Husserl's Origin of Geometry: An Introduction*, 86. Translation modified.
5. Derrida, *Speech and Phenomena*, 67.
6. See Derrida, "Freud and the Scene of Writing" and Derrida, *Of Grammatology*. It is worth recalling that "Freud and the Scene of Writing" was first published in 1966 and thus between the first appearance of *Of Grammatology*, in two successive issues of *Critique*, in 1965 and the final publication in a single volume in 1967.
7. Derrida, "Freud and the Scene of Writing," 250.
8. Ibid., 252.
9. Freud, *Project for a Scientific Psychology*, 295.
10. Ibid., 296.
11. Ibid.
12. The term "synapsis," used to account for what Freud calls "contact-barriers," will be introduced later, in 1897, by M. Foster and C. Scherrington.
13. Freud, *Project for a Scientific Psychology*, 299.

14. Ibid.

15. On the question of *Bahnung* and, more generally, on the *via rupta*, permit me to refer to F. Vitale, "Via rupta: vers la biodeconstruction."

16. Derrida, "Freud and the Scene of Writing," 252.

17. Freud, *Project for a Scientific Psychology*, 300.

18. Derrida, "Freud and the Scene of Writing," 254.

19. Ibid., 289: "The last part of the lecture concerned the arche-trace as erasure."

20. Ibid., 253. The importance of this passage is testified to by the fact that we find it almost literally reproduced in the programmatic essay "Différence" (1968). Cf. Derrida, "Différance," 18: "The concept of trace (*Spur*), or breaching (*Bahnung*), and of the forces of breaching, from the *Project* on, are inseparable from the concept of difference. The origin of memory, and of the psyche as (conscious or unconscious) memory in general, can be described only by taking into account the differences between breaches. Freud says so overtly. There is no breach without difference and no difference without trace. All the differences in the production of the unconscious traces and in the process of inscription (*Niederschrift*) can also be interpreted as moments of *différance*, in the sense of putting into reserve. According to a schema that never ceased to guide Freud's thought, the movement of the trace is described as an effort of life to protect itself by *deferring* the dangerous investment, by constituting a reserve (*Vorrat*)."

21. Derrida, "Freud and the Scene of Writing," 254. M. Hägglund develops an analogous reading that hinges on the analysis of différance as the condition of the spatio-temporal dialectics that necessarily affects all that is. Cf. Hägglund, *Radical Atheism: Derrida and the Time of Life*, 19: "On the one hand, the spacing of time has an ultratranscendental status because it is the condition for everything *all the way up* to and including the ideal itself. The spacing of time is the condition not only for everything that can be cognized and experienced, but also for everything that can be thought and desired. On the other hand, the spacing of time has an ultratranscendental status because it is the condition for everything *all the way down* to the minimal forms of life. As Derrida maintains, there is no limit to the generality of *différance* and the structure of the trace applies to all fields of the living." As regards arche-writing, in the context of a reading that privileges an engagement with Husserl over Freud, Hägglund affirms: "My reading of Husserl has sought to demonstrate the 'transcendental' necessity of such spatiotemporal inscriptions even for the most minimal synthesis or most elementary self-awareness. Arche-writing, however, is not only a transcendental condition for the experience of a finite consciousness; it is an 'ultratranscendental' condition for life in general" (ibid., 73). I remark that Hägglund uses "ultratranscendental" as a synonym of "quasi-transcendental," privileging the former for strategic reasons I cannot discuss here.

22. The reference in the explanatory note added at the opening of "Freud and the Scene of Writing" in *Writing and Difference* on the occasion of the book's

publication is even more explicit. See Derrida, "Freud and the Scene of Writing," 248: "The difference between the pleasure principle and the reality principle, for example, is not uniquely, nor primarily, a distinction, an exteriority, but rather the original possibility, within life, of the detour, of deferral (*Aufschub*) and the original possibility of the economy of death (cf. *Beyond the Pleasure Principle*, SE XVIII)."

23. Derrida, *Of Grammatology*, 9.

24. Ibid.

25. Ibid., 324. Derrida refers, without quoting it, to Wiener, *The Human Use of Human Beings: Cybernetics and Society* (1950). See Wiener, *The Human Use of Human Beings*, 31: "Here I want to interject the semantic point that such words as life, purpose, and soul are grossly inadequate to precise scientific thinking. These terms have gained their significance through our recognition of the unity of a certain group of phenomena, and do not in fact furnish us with any adequate basis to characterize this unity. Whenever we find a new phenomenon which partakes to some degree of the nature of those which we have already termed 'living phenomena,' but does not conform to all the associated aspects which define the term 'life,' we are faced with the problem whether to enlarge the word 'life' so as to include them, or to define it in a more restrictive way so as to exclude them. . . . Now that certain analogies of behavior are being observed between the machine and the living organism, the problem as to whether the machine is alive or not is, for our purposes, semantic and we are at liberty to answer it one way or the other as best suits our convenience. . . . It is in my opinion, therefore, best to avoid all question-begging epithets such as 'life,' 'soul,' 'vitalism,' and the like." See also ibid., 33: "While it is impossible to make any universal statements concerning life-imitating automata in a field which is growing as rapidly as that of automatization, there are some general features of these machines as they actually exist that I should like to emphasize. One is that they are machines to perform some definite task or tasks, and therefore must possess effector organs (analogous to arms and legs in human beings) with which such tasks can be performed. The second point is that they must be en rapport with the outer world by sense organs, such as photoelectric cells and thermometers, which not only tell them what the existing circumstances are, but enable them to record the performance or nonperformance of their own tasks." As we shall see, in the unpublished seminar *La vie la mort*, Derrida will discuss largely the cybernetics analogy between living being and machine. In a recent, well-documented essay, Christopher Johnson confirms that Norbert Wiener has been the main protagonist of cybernetics' propagation in French scientific culture. Cf. Johnson, "'French' Cybernetics."

26. On the notion of *différance* as "arche-synthesis," see Gasché, *The Tain of the Mirror*, 177–251.

27. For an interpretation of Derrida's reference to the information theory and the theory of systems and a first attempt to give account of their relationship with biology, see C. Johnson, *System and Writing in the Philosophy of Jacques Derrida*.

28. Cf. Debru et al., *Une nouvelle connaissance du vivant: François Jacob, André Lwoff, Jacques Monod*.

29. Cf. "Cahiers de Royaumont," Philosophie N° V, *Le concept d'information dans la science contemporaine*.

30. Cf. André Lwoff, "Le concept d'information dans la biologie moléculaire," 173–202.

31. Cf. G. Canguilhem, "Le concept et la vie," in *Revue philosophique de Louvain*, LXIV, 1966, published later in G. Canguilhem, *Etudes d'histoire et de philosophie des sciences* (Paris: Vrin, 1968) (see the English translation in G. Canguilhem, "The Concept of Life"). As we shall see in the seminar *La vie la mort*, Derrida discusses Canguilhem's essay by referring to the version included in the volume, in 1968, as if it followed the publication of *Of Grammatology*. On the relationship between Derrida and Canguilhem, I refer to Derrida's biography, especially where the reciprocal exchange of the respective writings since the early 1960s is attested to. As to where Derrida's affirmation according to which Canguilhem represented his "Philosophical superego (surmoi philosophique)" is recalled, see B. Peeters, *Derrida: A Biography*, 129.

32. Cf. Kay, "A Book of Life? How the Genoma Became an Information System and DNA a Language," 504–528; and Kay, *Who Wrote the Book of Life*, 307–315. The editors of Jacob's scientific writings ascribe to the latter an important role in the use of the scriptural metaphor in biology and in particular interpret this fact as the effect of the influence of structuralism in the France of the time. See M. Morange, "Introduction," 58: "The use of these metaphors is not due to Jacob but he was at the origin of some of them and, more than other biologists, put their value to the test in *The Logic of Life* as well as in articles and discussions with linguists. The French cultural context was certainly propitious to this development if one takes into account the importance of the French structuralist school and of Lévi-Strauss's work."

33. Derrida recalls this point in the first footnote of the *avertissement* that precedes the text, cf. *Of Grammatology*, 323.

34. Leroi-Gourhan, 370.

35. On Derrida's reference to Leroi-Gourhan apropos of the prehistoric genesis of the iterable trace and thus of the pre-cultural experience of spatio-temporality, permit me to refer to my "Via rupta: vers la biodeconstruction." For Derrida's recourse to the concept of "mytho-graphy" (that is, multidimensional symbolical writing), elaborated by Leroi-Gourhan in order to describe the evolutionary stage that goes from the first testimonies of writing (*graphie*) to writing in its restricted sense, I also recall my *Mitografie*.

36. Derrida, *Of Grammatology*, 84. Translation modified.

37. Leroi-Gourhan, *Gesture and Speech*, 219–236.

38. Ibid., 221.

39. Ibid., 222.

40. Ibid., 224.
41. Ibid., 328.
42. Derrida, *Of Grammatology*, 70.
43. Cf. Leroi-Gourhan, *Gesture and Speech*, 269: "In the first part of this book we often spoke about the evolution of the human body, the physical 'fund' of humanity. Having defined the two major criteria of technicity and language against the background of their common origin, we became aware of the close links between human evolution and a broader zoological context. The second part dealt principally with the collective entity of the cultural group. Technicity and language were viewed no longer in a zoological perspective but as phenomena subject to evolutionary laws which, although developing far more rapidly, run parallel to those of zoology. Human evolution was seen to be coherent in terms of its two fundamental characteristics of manual and verbal technicity although in a sense dissociated into two levels, that of phyletic evolution—as a result of which present-day peoples are a collection of individuals with physical properties hardly different from those possessed thirty thousand years ago—and that of ethnic evolution, which has turned humankind into an exteriorized body whose properties are globally in a state of accelerated change."
44. Derrida, *Of Grammatology*, 130. As is well known, Bernard Stiegler was the first to understand the importance of Leroi-Gourhan for Derrida, by developing in an original fashion the question of différance as the condition of human evolution and of the latter's development into a technical evolution. Cf. B. Stiegler, *Technics and Time, 1. The Fault of Epimetheus*, 136. "This analysis based on the work of Leroi-Gourhan will also allow for a dialogue with Jacques Derrida around the concept of différance, as this concept describes the process of life of which the human is a singular case, but only a case. What is in question is not emptying the human of all specificity but radically challenging the border between the animal and the human. . . . Since the *grammè* is older than the specifically human written forms, and because the letter is nothing without it, the conceptual unity that différance is contests the opposition animal/human and, in the same move, the opposition nature/culture. 'Intentional consciousness' finds the origin of its possibility before the human; it is nothing else but 'the emergence that has the *grammè* appearing as such.' *We are left with the question of determining what the conditions of such an emergence of the 'grammè as such' are, and the consequences as to the general history of life and/or of the grammè. This will be our question.*"
45. Derrida, *Of Grammatology*, 84.
46. In *On Beyond Living*, Richard Doyle refers to these passages from *Of Grammatology* by advancing an objection. He argues that Derrida would be fascinated by the introduction of the notion of "programme" in biology, so as to overlook its metaphysical matrix and deterministic bearings. In particular, Derrida would not understand that, in its application to biology, this notion implies a rigidly deterministic conception, the very dogma of modern biology according to

which the genetic code determines the logic of the living in an exclusive fashion, through a sequential process that is causal and thus linear. Cf. Doyle, *On Beyond Living*, 93: "With the appearance of new forms of writing such as the genetic code, Derrida argues, this logocentrism—the primacy of speech over writing—can be glimpsed as a metaphysics and not treated as a natural priority. Indeed, Derrida hopes to deconstruct the line that opposes nature and culture. And yet the news that Derrida receives of the 'genetic inscription' is of course itself written in lines. What Derrida perceives to be a sign of the end of the book and the beginning of writing is itself a sign of the 'line,' more specifically, the irreversible vectors in the central dogma of molecular biology. Here we find evidence in Derrida of the very flattening of 'life' inscribed by molecular biology, where the central dogma reads (even today, more or less) 'DNA makes proteins, proteins make us.' The line of power and information that flows from the gene to the 'body' is, under the central dogma, irreversible. DNA, in this account, is anointed with a style of sovereignty radically at odds with the Derridean account of writing without origins." And Doyle goes on: "Thus, what powers, fuels, or inflates the metaphysical moment of 'The Program' is a genetic, scientific account that metaphysically privileges precisely the linearity that Derrida critiques. This in itself is not surprising, nor does it necessarily call for critique. But the style of this inflation of 'The Program' into a sign not just of behavioral regulation, but, perhaps, of metaphysical necessity, calls for questioning. The alignment of the 'history of life' with the history of writing, to be sure, offers the possibility of disrupting the 'line' that leads from nucleic acids to 'us'—this is in fact one of the tasks of this book. But at the same time, by inscribing 'The Program,' however marginally and strategically, with the force of life and metaphysics, Derrida risks increasing the sovereignty of the 'Master Molecule,' DNA. Within the stability of the central dogma, 'genetic inscription' takes on the very onto-theological attributes deconstruction is meant to disrupt. This description of grammatology's implication in the central dogma is meant as more than merely an ironic marker of the interminable analysis of deconstruction, a snickering trickster catching Derrida at his own game. Rather, it is a trace of the force and momentum of the tropical alignment of DNA with language" (94). However, this criticism rests on a misunderstanding of Derrida's notion of "arche-writing," which, as we have seen, constitutes the condition of possibility of the inscription/iteration of a trace: the "gramma" is thus the form or elementary structure of the trace. The latter is structured according to the necessity of iteration, which does not necessarily entail the typical linearization of phonetico-alphabetic writing or the subordination of the trace to a meaning that precedes and is independent from it, but rather excludes that subordination. Therefore, Derrida is interested in the notion of "genetic programme" because it allows him to shed light on arche-writing as the condition of the genesis and structure of the life of the living. Ultimately, as we will see later, in the seminar *La Vie la mort*, Derrida elaborates an explicit deconstruction of the notion of genetic programme and thus of the deterministic dogma of modern biology.

47. This perspective is explicitly raised in Derrida, "Differance," 17: "The same, precisely, is *différance* (with an *a*) as the displaced and equivocal passage of one different thing to another, from one term of an opposition to the other. Thus one could reconsider all the pairs of opposites on which philosophy is constructed and on our discourse live, not in order to see opposition erase itself but to see what indicates that each of the terms must appear as the other different and deferred in the economy of the same (the intelligible as differing-deferring the sensible, as the sensible different and deferred; the concept as different and deferred, differing-deferring intuition; culture as nature different and deferred, differing-deferring; all the others of *physis*—*tekhnē*, *nomos*, *thesis*, society, freedom, history, mind, etc.,—as *physis* different and deferred, or as *physis* differing and deferring. *Physis* in *différance*."

48. Derrida, *Of Grammatology*, 9.

Chapter II

1. The quotations from the seminar indicate the number of the session and that of the page, according to the original draft.

2. Cf. Derrida, *Otobiographies*, 3–35.

3. Cf. Derrida, *Interpreting Signatures (Nietzsche/Heidegger): Two Questions*, 58–71.

4. On "proper name" and "signature," see the entries in G. Bennington, *Derridabase*, cit., 104–113 and 148–165. In particular 148: "My proper name outlives me. After my death, it will still be possible to name me and speak of me. Like every sign, including 'I,' the proper name involves the necessary possibility of functioning in my absence, of detaching itself from its bearer: and according to the logic we have already seen at work, one must be able to take this absence to a certain absolute, which we call death. So we shall say that even while I am alive, my name marks my death. It already bears the death of its bearer. It is already the name of the dead person, the anticipated memory of a departure. . . . The mark which identifies me, which makes me rather than anyone else, depropriates me immediately by announcing my death, separating me a priori from the same self it constitutes or secures. The signature, and this is precisely what distinguishes it from the proper name in general, attempts to catch up again the proper we have seen depropriate itself immediately in the name."

5. Derrida, *La vie la mort*, 1.1.

6. Cf. J. Derrida, *Positions*, 41.

7. Ibid., 42.

8. Cf. R. Esposito, "Community, Immunity, Biopolitics," 143: "We may say that Derrida, like Heidegger and Freud, thinks of life from the refraction angle of death; rather than thinking of death from the perspective of life," 149: "We may say that the fundamental divide of 20th century philosophy goes between those that think of life from the horizon of death, like Heidegger, Freud and Derrida

himself, and those, like Nietzsche, Bergson and Foucault, who think of death from the horizon of life." Hence biopolitics limits itself to the mere overturning of the hierarchy and thus leaves unaltered the system of opposition and the metaphysics that grounds this system, whereas deconstruction aims to think beyond or before this system the différance between life and death. On this point, see M. Hägglund, *Radical Atheism*, 48: "Derrida proposes neither a philosophy of life nor a philosophy of death but insists on the stricture of 'life-death.'"

9. Derrida, *La vie la mort*, 1.2.
10. Ibid., 1.2.
11. Cf. Jacob, *The Logic of Life*, 251: "According to Norbert Wiener, there is no obstacle to using a metaphor 'in which the organism is seen as a message.' . . . With the possibility of carrying out mechanically a series of operations laid down in a programme, the old problem of the relations between animal and machine was posed in new terms. 'Both systems are precisely parallel in their analogous attempts to control entropy through feedback,' said Wiener. Both succeed by disorganizing the external environment, 'by consuming negative entropy,' to use the expression of Schrödinger and Brillouin. Both have special equipment, in fact, for collecting at a low energy-level the information coming from the outside world and for transforming it for their own purposes." For the quotations from Wiener, see N. Wiener, *The Human Use of Human Beings: Cybernetics and Society*, 95, 26.
12. On the occasion of the publication of Jacob's book, Michel Foucault wrote an article titled "Croître et multiplier," which appeared in *Le Monde* in 1970. Here he acknowledges the revolutionary significance of this book. Cf. Foucault, "Croître et multiplier," 99: "François Jacob has just written a truly great book of history. He does not tell us how the laws and mechanism of heredity have been discovered step by step, but what genetics has revolutionized in the oldest knowledge of the West. This noteworthy book tells us how and why we must think life, time, individuals, and randomness, otherwise. And this does not happen at the borders of the world, but here, in the little machinery of our cells." However, for Foucault, the revolution introduced by the genetic programme is a threat to be conjured away in order to defend a rather indeterminate "secret" of life from the deterministic power of calculation that this programme seems to make possible, cf. ibid., 103: "Does this mean the return to the animal-machine, the triumph of the existence-fermentation, since the mysterious specificity of life has been deleted? This question makes no sense now, and yet one can say to what extent the cell is a measure of physico-chemical reactions, to what extent it works as a calculator. It is the notion of programme that is now at the centre of biology. A biology without life? . . . One should not dream about life as a great and continuous creation of individuals; one should think of the living as the calculable play of randomness and reproduction. Jacob's book is the most remarkable history of biology that has ever been written, but it also invites us to a great re-education of thought [à un grand réapprentissage de la pensée]. *The Logic of Life* shows what science needed to

know at the same time as what this knowledge cost to thought." However, if the secret of life that must be defended against the calculating abstraction of science is the secret of creation, or creation as a secret, and thus a creation that is detached from the possibility of men, then Foucault's position is more or less consciously compromised by a metaphysico-religious presupposition, a presupposition betrayed from the title itself that refers to the passage from *Genesis* concerning the creation of man and woman. Cf. *Genesis*, 1, 26–29: "26. And God said, Let us make man in our image, after our likeness: and let them have dominion over the fish of the sea, and over the fowl of the air, and over the cattle, and over all the earth, and over every creeping thing that creepeth upon the earth. 27. So God created man in his own image, in the image of God created he him; male and female created he them. 28. And God blessed them, and God said unto them, *Be fruitful, and multiply* [my emphasis], and replenish the earth, and subdue it: and have dominion over the fish of the sea, and over the fowl of the air, and over every living thing that moveth upon the earth. 29. And God said, Behold, I have given you every herb bearing seed, which is upon the face of all the earth, and every tree, in the which is the fruit of a tree yielding seed; to you it shall be for meat." The French translation of the passage says: "Et Dieu les bénit, et leur dit: *Croissez, et multipliez*, et remplissez la terre; et l'assujettissez, et dominez sur les poissons de la mer, et sur les oiseaux des cieux, et sur toute bête qui se meut sur la terre" (David Martin, 1744, *Genèse*, I: 28). As we will see, in his reading of Jacob, Derrida has no interest in this kind of humanistic, if not religious, arguments, and he makes the most for deconstruction precisely from what scares Foucault most: the structural priority of repetition as the condition of the life of the living. In fact, for Foucault, repetition or repeatability constitutes the most serious threat to the secret of life understood, as we have seen, as the continuous creation of the individual. If there is repetition, and not the individual, at the origins of life, then one cannot think the unconditioned and unrepeatable unity of creation. Cf. Foucault, "Croître et multiplier," 100: "Genetics also strikes us in several other ways; it touches upon some of the fundamental postulates, on which our transitory truths are formed and some of our eternal dreams are gathered together. Jacob's book calls them into question. I will be content with one of the best-anchored dreams: the one that subordinates reproduction to the individual, to its growth and is death. For long time, we believed that reproducing was for the individual at the end of its growth a means for prolonging itself beyond itself, for compensating death by transferring in the future this duplication far from its form. It has taken us fifty years to know that the metabolism of the cell and the mechanisms of the growth of the individual are governed by a code deposed in the DNA of the nucleus and transmitted through messengers, to know thus that the whole, little chemical laboratory of a bacterium is bound to produce a second one (that is its dream, Jacob says), to know that the most complex forms of organization (alongside sexuality, death, its partner, signs and language, their remote effects) are mere detours to assure once more that the reproduction takes place . . . The bacterium: a reproduction machine, which reproduces its mechanism of reproduction, a material of heredity that indefinitely proliferates by itself, a mere repetition that precedes the singularity of the individual. Across its evolution, the living was a duplicating machine much earlier than an individual organism."

13. Jacob, *The Logic of Life*, 254.
14. Ibid., 264.
15. Cf. Jacob, *The Logic of Life*, 299: "the operational value of the concept of life has continually dwindled and its power of abstraction declined. Biologists no longer study life today. They no longer attempt to define it. Instead, they investigate the structure of living systems, their functions, their history."
16. Derrida, *La vie la mort*, 1.1.
17. Ibid., 1.3.
18. Ibid., 1.1.
19. Derrida, *Glas*, 83.
20. Hegel, *Philosophy of Nature*, 212. The significance of the Phoenix as a *religious* symbol is discussed in Hegel's *Lectures on the Philosophy of Religion*, vol. II, 83–84. See also *The Philosophy of History*, 72–73: "But the next consideration which allies itself with that of change, is, that change while it imports dissolution, involves at the same time the rise of a *new* life—that while death is the issue of life, life is also the issue of death. This is a grand conception . . . the *Phoenix* as a type of the Life of Nature; eternally preparing for itself its funeral pile, and consuming itself upon it; but so that from its ashes is produced the new, renovated, fresh life."
21. For the figure of phoenix in Hegel, see Derrida, *Glas*, 102, 103, 116, 117, 155.
22. See A. Kojève, *Introduction à la lecture de Hegel*, 537: "The Man 'appears' as a being that is always conscious of its death, often accepts it freely and, sometimes, gives it to himself voluntarily. Therefore, the 'dialectical' or anthropological philosophy of Hegel is, ultimately, *a philosophy of death*."
23. Derrida, "From Restricted to General Economy," 323.
24. Ibid., 327–328.
25. Derrida, *Glas*, 139.
26. Ibid., 141.
27. Hegel, *Science of Logic*, 687.
28. Hegel, *Philosophy of Nature*, 9.
29. Hegel, *The Philosophy of History*, 95.
30. Derrida, *Glas*, 28.
31. Ibid., 31.
32. Hegel, *The Spirit of Christianity*, 260.
33. Hegel, *Philosophy of Nature*, 313.
34. Ibid., 323–324.
35. Cf. Hartsoeker, *Essay de dioptrique*.

36. Cf. Swammerdam, *Miraculum naturae sive uteri muliebris factorya*.
37. See Diderot and d'Alemebrt, *Encyclopédie*, vol. VII: "Génération," 563ff.
38. Derrida, *Force and Signifcation*, 26.
39. Hegel, *Philosophy of Nature*, 55.
40. Derrida, *Glas*, 114.
41. Hegel, *Philosophy of Nature*, 345.
42. Ibid., 441.
43. Derrida, *Glas*, 116.

Chapter III

1. Derrida, *La vie la mort*, 1.22.
2. Canguilhem, "The Concept of Life," 316.
3. Derrida, *La vie la mort*, 1.20.
4. Canguilhem, "The Concept of Life," 319.
5. Jacob, *The Logic of Life*, 1.
6. Ibid., 20.
7. Ibid., 2.
8. Ibid., 8.
9. Also, Jacques Monod insists on the necessity of liberating teleology from its metaphysical matrix of Aristotelian derivation by reformulating this notion in a mechanistic perspective. He introduces the concept of "teleonomy" in order to indicate the totality of those biological processes that are teleologically oriented to the realization, conservation, and transmission of the information retained in the genetic programme. Therefore, for Monod, as well as for Jacob, it is enough to place the genetic programme at the origins of teleonomy in order to free the latter from its metaphysico-theological roots, although Monod also admits that the genetic programme itself could respond to a more general purposiveness of nature. See Monod, *Chance and Necessity*, 21: "The cornerstone of the scientific method is the postulate that nature is objective. In other words, the *systematic* denial that 'true' knowledge can be got at by interpreting phenomena in terms of final causes—that is to say, of 'purpose.' An exact date may be given for the discovery of this canon. The formulation by Galileo and Descartes of the principle of inertia laid the groundwork not only for mechanics but for the epistemology of modern science, by abolishing Aristotelian physics and cosmology. To be sure, neither reason, nor logic, nor observation, nor even the idea of their systematic confrontation had been ignored by Descartes' predecessors. But science as we understand it today could not have been developed upon those foundations alone. It required the unbending stricture implicit in the postulate of objectivity—ironclad, pure, forever undemonstrable. For it is obviously impossible to imagine an experiment which could prove the *nonexistence* anywhere in nature of a purpose, of a pursued

end. But the postulate of objectivity is consubstantial with science; it has guided the whole of its prodigious development for three centuries. There is no way to be rid of it, even tentatively or in a limited area, without departing from the domain of science itself. Objectivity nevertheless obliges us to recognize the teleonomic character of living organisms, to admit that in their structure and performance they act projectively—realize and pursue a purpose." It is worth noting that Monod does not question the teleologico-metaphysical legacy of Cartesian mechanicism, to begin with, the distinction between *res cogitans* and *res extensa*, and in particular the principle of the immutability of the divine will that, for Descartes, constitutes the ultimate ground for the interpretation of nature as a machine. On this point, permit me to refer to Vitale, "With or Without You . . . Deconstructing Teleology between Philosophy and Biology."

10. Derrida, *La vie la mort*, 1.8.
11. Jacob, *The Logic of Life*, 2.
12. Ibid.
13. Ibid., 2.
14. Derrida, *La vie la mort*, 1.15.
15. For an accurate analysis of the deconstruction of analogy and of its effects on both philosophy and literary studies, see Gasché, *The Tain of the Mirror*, 293–318. In particular, cf. 304: "As Derrida has demonstrated in *Plato's Pharmacy*, a certain dominating and decisive hierarchization takes place between the terms of the relations that enter into correspondence in a relation of analogy. This hierarchizing authority of logocentric analogy comes from the fact that one term within the relation of relations comes to name the relation itself. Consequently, all the elements that make up the relations find themselves comprised by the structure that names the relation of analogy as a whole. That name, ultimately, is that of the logos." See also the entry on "Metaphor" in J. Bennington, *Derridabase*, 119–132.
16. Derrida, *La vie la mort*, 1.15.
17. Ibid., 1.16.
18. Ibid., 1.17.
19. Ibid.
20. Derrida, "Différance," 18.
21. Derrida, *For What Tomorrow*, 39.
22. Ibid.
23. Ibid., 206n.
24. Jacob, *The Logic of Life*, 3.
25. Ibid.
26. Ibid., 292.
27. Derrida, *La vie la mort*, 1.19.
28. For a deeper engagement with epigenetics, see Jablonka and Lam, *Evolution in Four Dimensions*.
29. Francis, *Epigenetics: The Ultimate Mystery of Inheritance*, 125.

30. Ibid., 133.

31. Ibid., 159.

32. See Malabou, *Plasticity at the Dusk of Writing*, 57–59: "The constitution of writing as a motor scheme was the result of a gradual movement that began with structuralism and found its mooring in linguistics, genetics, and cybernetics. A pure linguistic image, the image of the gap or difference gradually established itself as the scheme of an ontological organization. The impact of a book such as François Jacob's *The Logic of Life* extended, even if not explicitly, Lévi-Strauss conclusions about linguistics: it is the theoretical expression of a certain organization of the real, a global morphology made up of 'meaningful gaps' and differences. *The Logic of Life* confirmed the existence of this linguistic structure of being by privileging the role of writing within it . . . In this graph ontology, the origin—whatever meaning is attributed to this word—can only be thought in terms of a trace, that is, a difference to the self. Generally, it is the concept of *program*—which is obviously also a concept in the field of cybernetics—that culminates and completes the constitution of the graphic scheme as the motor scheme of thought. Derrida alone recognized the full importance of this fulfillment and culmination . . . Derrida describes here [at the opening of *Of Grammatology*] the *semantic enlargement* of the concept of writing, not as an arbitrary philosophical decision but as an event, the appearance of a new order, starting from the pregnancy of the motifs of program, information or code. It is only on the basis of this programmatic organization of the real as it is liable to come to the awareness of an era that writing was able to constitute itself as a *philosophical* motor scheme." See also C. Malabou, "The End of Writing. Grammatology and Plasticity," 431–441. Furthermore, there would be much to say about the concept of motor scheme on which Malabou's whole argument rests. This concept implies a rather relaxed recourse to the classical opposition between the sensible and the intelligible and to the classical motive of imagination as a middle term. Despite the vicissitudes of the supposed graphic paradigm, the motor schema has no hold on deconstruction, for which it is structurally unsustainable.

33. Derrida, *Dissemination*, 52. See also Derrida, *Otobiographies*, 29–30. On these passages, see Senatore, *Of Seminal Difference*, 67–71.

34. Derrida, *La vie la mort*, 1.20.

35. Ibid., 1.9. Among the modern biologists of his time who support the emancipation of biology from philosophy, Derrida refers to Jacques Monod, whose position on this point would be more radical than Jacob's. See Derrida, *La vie la mort*, 1.9. This is the only place Monod is mentioned. Cf. J. Monod, *Chance and Necessity*.

36. In the seminar sessions devoted to Heidegger's critique of Nietzsche's alleged biologism, Derrida explicitly highlights the metaphysical matrix of this presupposition: cf. Derrida, *La vie la mort*, 10.11: "In its greatest formality, this scheme is the following and I believe that where, in reading Nietzsche, Heidegger aims to take a step forward or to take it after a step forward beyond metaphysics,

this scheme, at that precise moment, is the most traditional metaphysical scheme, the most Hegelian one, also for what concerns the relationship between science and philosophy. . . . the sciences are particular, they deal with a determinate type of being [*étant*] (or of object, as we may say with Kant or Husserl, being the object a determination of being [*étant*]). The sciences are thus regional, they do not deal but with a region of the objectivity of being. . . . When the properly scientific work of physics, mathematics or history begins, there the scholar as such must be sure of the sense of being, of the ontic region or the region of objectivity in which he works, in order to present himself as a physician, a mathematician or an historian. The scholar deals with the being or the object but neither thinks [*pense*] of the determinate beingness [*étantité*] or objectivity of his object, nor, a fortiori, of beingness or objectivity in general or of the totality of being or objectivity. It is philosophy (or metaphysics) that raises the question of knowing what is the physical being as such and a fortiori the being in its totality. It is philosophy that distributes and assigns to the regional sciences the senses of their field."

37. Derrida, *La vie la mort*, 1.9.
38. Ibid., 1.10.
39. Ibid.
40. Derrida, *La vie la mort*, 4.6.
41. Jacob, *The Logic of Life*, 306.
42. Ibid., 299.
43. Derrida, *La vie la mort*, 4.8.
44. Ibid.
45. Ibid., 4.7.
46. Jacob, *The Logic of Life*, 306.
47. Derrida, *La vie la mort*, 4.8.
48. Hegel, *Science of Logic*, 683.
49. Jacob, *The Logic of Life*, 2.
50. Ibid., 4.
51. Ibid., 5.
52. Ibid., 17.
53. Aristotle, *On the Parts of Animals*, 6.
54. Ibid., 16.
55. Derrida, *La vie la mort*, 4.12.
56. Ibid.
57. The same criticism can be addressed to Monod, who maintains that he can free teleology from its metaphysical roots by reformulating it in terms of "teleonomy" and thus subordinating the latter to the execution of the programme of invariance contained in the DNA. To put it shortly, Monod explains that three properties characterize the living by demarcating it from the other beings, both natural and artificial, namely "teleonomy, autonomous morphogenesis, and reproductive invariance" (Monod, *Chance and Necessity*, 13). The autonomous morphogenesis is "a

structure giving proof of an autonomous determinism" (ibid., 10), says Monod, by referring to the ability of the living to produce its own structure by itself, according to an internal principle. "Reproductive invariance" or, merely, "invariance," refers to the "hability to reproduce and to transmit *ne varietur* the information corresponding to their own structure" (ibid., 12). Therefore, teleonomy is the property "of being *objects endowed with a purpose or project,* which at the same time they exhibit in their structure and carry out through their performances. Rather than reject this idea (as certain biologists had tried to do) it is indispensable to recognize that it is essential to the very definition of living beings" (ibid., 9). However, we cannot see what distinguishes these properties: they are all teleological manifestations: self-production and invariant transmission are evidently understandable and definable only from the result that they produce and thus as processes oriented to an end. After all, it is Monod himself who reduces the genesis and structure of the living to the realization of the essential teleonomic project, even if he acknowledges the ambiguity of the notion of project: "That in turn will enable us to bring into better focus the notion most immediately and plainly inspired by the examination of the structures and performances of living beings, that of teleonomy. Analysis, nevertheless reveals it to be a profoundly ambiguous concept, since it implies the subjective idea of a 'project.' . . . But it is only as a part of a more comprehensive project that each individual project, whatever it may be, has any meaning. All the functional adaptations in living beings fulfill particular projects which may be seen as so many aspects or fragments of a unique and primary project, which is the preservation and multiplication of the species. To be more precise, we shall arbitrarily choose to define the essential teleonomic project as consisting in the transmission from generation to generation of the invariance content of characteristic of the species" (ibid., 14). It is worth noting that the definition of this essential teleonomic project does not only imply, through the notion of project, the reference to a teleological intentionality. It also implies the subordination of the finality that can be found in the process of the autonomous morphogenesis of the living to a more general and fundamental one: "that of the conservation and multiplication of the species." Now, even if this subordination does not necessarily entail the reference to a supernatural intention or intelligence, a personal God, the argumentative structure provided by Monod reproduces the metaphysical scheme of classical onto-theology. Among the particular teleonomic phenomena that can be observed in nature, Monod picks up that of the conservation and multiplication of the species in order to hypostatize it as an essential finality of the living in general, the ultimate finality, the essence of life. As Derrida taught us, we should recognize this schema as the determination of the "transcendental meaning" within the horizon of classical metaphysics; see for instance Derrida, *Positions,* 29: "At the point at which the concept of *differance,* and the chain attached to it, intervenes, all the conceptual oppositions of metaphysics (signifier/signified; sensible/intelligible; writing/speech; passivity/activity; etc.)—to the extent that they ultimately refer to the presence of something present (for example, in the form of the identity of the subject who is present for all his operations, present beneath every accident or event, self-present in its 'living speech,' in its enunciations, in the present objects and acts of its language, etc.)—become nonpertinent. They all amount, at one moment or another, to a subordination of the movement of *differance* in favor of the presence of a value or a *meaning* supposedly antecedent to *differance*, more original than it, exceeding and governing it in the last analysis. This is still the presence of what we called above the 'transcendental signified.'" For a further elaboration of these points, see Vitale, "With or Without You . . . Deconstructing Teleology between Philosophy and Biology."

58. Ibid., 4.14.
59. Ibid., 4.7.
60. Ibid., 5.2.
61. See Derrida, *La vie la mort*, 5.3: "I say 'historial' as it is not a historical question, one of the questions of or within history, but a question that concerns the historical as such, as it is determined as such according to what is in question here, insofar as history itself in its own historicity is related to the question of producing."
62. Ibid.
63. Cf. Derrida, *Dissemination*, 6: "There is no such thing as a 'metaphysical-concept.' There is no such thing as a 'metaphysical-name.' The 'metaphysical' is a certain determination or direction taken by a sequence or 'chain.' It cannot as such be opposed by a concept but rather by a process of textual labor and a different sort of articulation."
64. Derrida, *La vie la mort*, 5.4.
65. Ibid., 5.9.
66. Ibid., 5.1.
67. Ibid.
68. Ibid., 5.2.
69. Derrida, "Freud and the Scene of Writing," 254.
70. Derrida, *La vie la mort*, 5.3.
71. Jacob, *The Logic of Life*, 309.
72. Ibid., 297.
73. Derrida, *Dissemination*, 103: "In order for these contrary values (good/evil, true/false, essence/appearance, inside/outside, etc.) to be in opposition, each of the terms must be simply *external* to the other, which means that one of these oppositions (the opposition between inside and outside) must be already accredited as the matrix of all possible opposition. And one of the elements of the system (or of the series) must also stand as the very possibility of systematic or seriality in general."
74. Derrida, *La vie la mort*, 5.13.
75. Ibid., 5.15.
76. Jacob, *The Logic of Life*, 309.
77. Derrida, *La vie la mort*, 5.16.

78. Ibid.
79. Ibid., 5.17.
80. Jacob, *The Logic of Life*, 291.
81. Derrida, *La vie la mort*, 5.18.
82. Jacob, *The Logic of Life*, 309.
83. Ibid., 293.
84. Narra and Ocham, "Of What Use Is Sex to Bacteria?," 705. For an excellent review on these mechanisms, see Nielsen and Thomas, "Mechanisms of, and Barriers to, Horizontal Gene Transfer between Bacteria," 711–721.
85. Narra and Ocham, "Of What Use Is Sex to Bacteria?," 705: "Free-living bacteria reproduce by binary fission, whereby a single cell replicates its genome and divides into two identical daughter cells. In this system, variation is introduced into a lineage largely by mutations and sometimes by the intragenomic duplication of the existing genetic information. When limited to these processes, the generation of variation at the DNA level and the origin and spread of novel traits at the phenotypic level can be relatively slow processes in asexually reproducing lineages. Given that most mutations of any consequence to the organism (i.e., non-neutral mutations) are harmful, asexual lineages also suffer from the accumulation of deleterious mutations. . . . Naturally, both sexual reproduction (in which alleles from different organisms are merged) as well as recombination bring together new combinations of alleles over loci in each generation and can help overcome some of the drawbacks associated with asexuality. On the other hand, sexual reproduction disrupts combinations of unlinked beneficial mutations and, in such cases, asexual reproduction, in producing uniform progeny, would be analogous to winning a lottery."
86. Ibid., 708. The entire passage says: "Now that genome sequences of several *E. coli* strains have been resolved, the large differences in genome size observed amongst these strains have been shown to be the result of gene acquisition, mostly mediated by bacteriophages. In fact, lateral gene transfer has played such a prominent role in the content of bacterial genomes that it appears that vertically transmitted genes (i.e., those genes common to all strains) make up a minor portion of the entire *E. coli* gene pool, and that the majority of genes in all bacterial genomes were acquired laterally at some time during the evolutionary history of the lineage."
87. Ibid., 706.
88. Jacob, *The Logic of Life*, 273: "Each chemical species is reproduced exactly from one generation to another. Each chemical species does not, however, form copies of itself. A protein is not born of an identical protein. Proteins do not reproduce. They are organized from another substance, deoxyribonucleic acid, the constituent of chromosomes. This compound is the only one in the cell that can be reproduced by copying itself."
89. Ibid., 274.
90. Derrida, *La vie la mort*, 4.19.

Chapter IV

1. Jacob, *The Logic of Life*, 275.
2. For the preeminence of the syntactic dimension over the semantic in Derrida's elaboration of synthetic infrastructures, see Gasché, *The Tain of the Mirror*, 239–251.
3. Jacob, *The Logic of Life*, 306.
4. Canguilhem, "The Concept of Life," 362.
5. Jacob, *The Logic of Life*, 305.
6. Derrida, *La vie la mort*, 6.2.
7. Ibid.
8. Derrida, *Of Grammatology*, 158.
9. Derrida, *La vie la mort*, 4.1.
10. Derrida, "Living On," 83. See also J. Derrida, *Dissemination*, 35–36; and, in particular, Derrida, *Limited Inc.*, 148: "I wanted to recall that the concept of text I propose is limited neither to the graphic, nor to the book, nor even to discourse, and even less to the semantic, representational, symbolic, ideal, or ideological sphere. What I call 'text' implies all the structures called 'real,' 'economic,' 'historical,' socio–institutional, in short: all possible referents. Another way of recalling once again that 'there is nothing outside the text.' That does not mean that all referents are suspended, denied, or enclosed in a book, as people have claimed, or have been naive enough to believe and to have accused me of believing. But it does mean that every referent, all reality has the structure of a differential trace, and that one cannot refer to this 'real' except in an interpretive experience. The latter neither yields meaning nor assumes it except in a movement of differential referring. That's all."
11. Derrida, *La vie la mort*, 6.17.
12. Ibid., 6.5.
13. Ibid., 6.4.
14. Ibid., 4.15.
15. Jacob, *The Logic of Life*, 14.
16. Derrida, *La vie la mort*, 6.6.
17. Ibid.
18. It is from this perspective that a dialogue with biosemiotics could be opened. In this area of research, which is quite young and yet already differentiated, living processes, from DNA to the relationship with the environment, are described as semiotic processes. Therefore, biosemiotics aims to overcome the cybernetic paradigm, in particular, its rigid determinism, in order to account for the complexity of living processes. In particular, we may consider a dialogue with the Piercean line of biosemiotics, which goes back to Thomas A. Sebeok, who is generally recognized as the founder and promotor of this area of research. Sebeok was a student of Yuri Lotman and Roman Jakobson: he combines Peirce's semiot-

ics with the theoretical biology of Jacob von Uexküll, taking as his starting point the essential role that both assign to interpretation, the one in the achievement of semiotic processes, the other in the relationship that the living entertains with its environment (*Umwelt*) (cf. Sebeok, "Biosemiotics: Its Roots, Proliferation, and Prospects," Barbieri, *Introduction to Biosemiotics*; and Favareau, *Essential Reading in Biosemiotics*). As is well known, Derrida recognized a privilege to Peirce's semiotics with respect to the linguistic of Saussurre, precisely because of the role that Peirce ascribes to interpretation. Cf. Derrida, *Of Grammatology*, 49: "Peirce goes very far in the direction that I have called the de-construction of the transcendental signified, which, at one time or another, would place a reassuring end to the reference from sign to sign. I have identified logocentrism and the metaphysics of presence as the exigent, powerful, systematic, and irrepressible desire for such a signified. Now Peirce considers the indefiniteness of reference as the criterion that allows us to recognize that we are indeed dealing with a system of signs. What broaches the movement of signification is what makes its interruption impossible. The thing itself is a sign." In particular, we could open a dialogue with the work of Tuomo Jämsä, one of the most recognized representatives of biosemiotics, who starts from a re-elaboration of the Piercean model of semiosis that inspired Sebeok in order to extend it to the interpretation of the elementary processes of life and evolution. Cf. Jämsä, "Semiosis in Evolution," 72: "The term 'interpreting subject' in the graph [Peircean model of semiosis] gives to understand that the biosemiotic premises hold true for the communication between an organism and its habitat. Jakob von Uexküll has introduced the name 'Umwelt' to represent the world an organism lives in. The life can be described as a continuous dialogue in the function cycle between an interpreting subject and its umwelt. Interpretation is composed of 'writing and reading,' of encoding meanings into signs and decoding signs into meanings. From the point of view of a reader not familiar with the Peircean doctrine of signs, it is necessary to point out that signs are not only morphemes (meaningful items) of language or paralinguistic traits of human or animal communication but also patterns or things in the umwelt called icons (images or the like) and indices (all kinds of entities inside and outside an organism)." From this perspective, it is not by chance that Jämsä acknowledges the necessity of abandoning the linguistic model in favor of a textual one. Cf. 93: "Decisive for biosemiosis is what I would call the 'textual component.' Kalevi Kull has with reason highlighted the precedence of the text—the 'biotext,' as he labels it—over the sign. There is no deeper polarity between the two, however. Naming the sign processes 'texts' underscores the presence of the semiotic grammar, which is so closely alike to that of language. The biotexts in the bodies of organisms differ some from language texts because the propositions they are composed of are often simultaneous and the repetition makes them more redundant. The term 'text' is factually more relevant and gives a better understanding of the unbelievable complexity of the bodily text. Our knowledge of the biotext is imperfect, with many lacunae, so far. The bodily text compares to the notions of text and utterance in speech act theory. In it, 'illocutionary force' is the basic intentional meaning what a text or utterance has. From this point of view, the illocutionary force of a biotext is to keep a certain body alive. The statement brings us into the domain of the pragmatic semiotic component. Survival is the highest pragmatic meaning." See also Kull, "A sign is not alive—a text Is."

19. Ibid.

20. As is well known, it was the physicist Erwin Schrödinger in *What is Life?* who applied to the study of life the notion of entropy as the mechanico-statistic measure of the disorder of the molecular state of a system. It is worth recalling that the concept of entropy is introduced at the beginning of XIX century, in the field of thermodynamics, in order to describe a characteristic shared by the systems known until then (a characteristic whose generality was observed first by Carnot in 1824), that is, the fact that transformations happen spontaneously only in one direction, that of a greater disorder. In particular, the word "entropy" was first introduced by Rudolf Clausius in his *Abhandlungen über die mechanische Wärmetheorie* (treatise on the mechanical theory of heat), published in 1864. Clausius referred properly to the link between the movement internal to a body or a system and internal energy or heat, a link that made explicit the great intuition of Enlightenment, that the heat was related to the mechanical movement of particles internal to the body. Schrödinger has recourse to the mechanico-statistic notion of entropy in order to explain the process of metabolism. He thinks that the aim of metabolism is not the assimilation of material substances from the organism and that it would be absurd to claim that, through metabolism, the organism gains energy. Cf. Schrödinger, *What Is Life?*, 70: "How does the living organism avoid decay? The obvious answer is: by eating, drinking, breathing and (in the case of plants) assimilating. The technical term is *metabolism*. The Greek word means change or exchange. Exchange of what? Originally the underlying idea is, no doubt, exchange of material. (E.g. the German for metabolism is *Stoffwechsel*.) That the exchange of material should be the essential thing is absurd. Any atom of nitrogen, oxygen, sulphur, etc., is as good as any other of its kind; what could be gained by exchanging them? For a while in the past our curiosity was silenced by being told that we feed upon energy. In some very advanced country (I don't remember whether it was Germany or the U.S.A. or both) you could find menu cards in restaurants indicating, in addition to the price, the energy content of every dish. Needless to say, taken literally, this is just as absurd. For an adult organism the energy content is as stationary as the material content. Since, surely, any calorie is worth as much as any other calorie, one cannot see how a mere exchange could help. What then is that precious something contained in our food which keeps us from death?" For Schrödinger, the aim of metabolism is assimilating order from the external environment in order to contrast the tendency to disorder: "Thus the device by which an organism maintains itself stationary at a fairly high level of orderliness (= fairly low level of entropy) really consists in continually sucking orderliness from its environment. This conclusion is

less paradoxical than it appears at first sight. Rather could it be blamed for triviality. Indeed, in the case of higher animals we know the kind of orderliness they feed upon well enough, viz. the extremely well-ordered state of matter in more or less complicated organic compounds, which serve them as foodstuffs. After utilizing it they return it in a very much degraded form—not entirely degraded, however, for plants can still make use of it. (These, of course, have their most powerful supply of 'negative entropy' in the sunlight)." This conception of metabolism as a system to assimilate order from outside will open the way to cybernetics, which translates it into information, and to its application in biology, as Jacob himself points out. Cf. Jacob, *The Logic of Life*, 249: "Statistical mechanics made it possible to interpret the average behaviour of large populations of molecules. Genetic analysis, however, revealed that biological properties were not the result of statistical molecular events; but that, instead, they were based on the quality of some substances contained in the chromosomes. In contrast to the order of inanimate bodies, the order of living organisms could not be extracted from disorder. It depends on the reproduction of an already existing order. According to Schrödinger, 'Life seems to be orderly and lawful behaviour of matter, not based exclusively on its tendency to go over from order to disorder, but based partly on existing order that is kept up.' In the middle of the nineteenth century the concept of information opened the way to the investigation and transmission of this order." Cf. Wiener, *The Human Use of Human Beings*, 39: "It is my thesis that the physical functioning of the living individual and the operation of some of the newer communication machines are precisely parallel in their analogous attempts to control entropy through feedback. Both of them have sensory receptors as one stage in their cycle of operation: that is, in both of them there exists a special apparatus for collecting information from the outer world at low energy levels, and for making it available in the operation of the individual or of the machine. In both cases these external messages are not taken neat, but through the internal transforming powers of the apparatus, whether it be alive or dead. The information is then turned into a new form available for the further stages of performance."

21. Jacob, *The Logic of Life*, 250.
22. Ibid., 251.
23. Derrida, *La vie la mort*, 6.8.
24. Ibid.
25. Ibid., 6.9. Also in this case, Wiener is the source of this reduction and subordination of energy with respect to information. However, his argument appears specious, as it merely rests on the example of the photosynthesis of plants. Cf. Wiener, *The Human Use of Human Beings*, 38: "Quantum theory has led, for our purposes, to a new association of energy and information. A crude form of this association occurs in the theories of line noise in a telephone circuit or an amplifier. Such background noise may be shown to be unavoidable, as it depends on the discrete character of the electrons which carry the current; and yet it has a

definite power of destroying information. The circuit therefore demands a certain amount of communication power in order that the message may not be swamped by its own energy. More fundamental than this example is the fact that light itself has an atomic structure, and that light of a given frequency is radiated in lumps which are known as light quanta, which have a determined energy dependent on that frequency. Thus there can be no radiation of less energy than a single light quantum. The transfer of information cannot take place without a certain expenditure of energy, so that there is no sharp boundary between energetic coupling and informational coupling. Nevertheless, for most practical purposes, a light quantum is a very small thing; and the amount of energy transfer which is necessary for an effective informational coupling is quite small. It follows that in considering such a local process as the growth of a tree or of a human being, which depends directly or indirectly on radiation from the sun, an enormous local decrease in entropy may be associated with quite a moderate energy transfer." It is worth recalling that Schrödinger, in an appendix to the aforementioned chapter on entropy, admits that he was wrong in reducing the function of metabolism to a mere assimilation of order, and thus acknowledges the necessity of taking into account the assimilation of energy as its condition. Cf. Schrödinger, *What Is Life?*, 74: "The remarks on negative entropy have met with doubt and opposition from physicist colleagues. Let me say first, that if I had been catering for them alone I should have let the discussion turn on free energy instead. It is the more familiar notion in this context. But this highly technical term seemed linguistically too near to energy for making the average reader alive to the contrast between the two things. He is likely to take free as more or less an *epitheton ornans* without much relevance, while actually the concept is a rather intricate one, whose relation to Boltzmann's order-disorder principle is less easy to trace than for entropy and 'entropy taken with a negative sign,' which by the way is not my invention. It happens to be precisely the thing on which Boltzmann's original argument turned. But F. Simon has very pertinently pointed out to me that my simple thermodynamical considerations cannot account for our having to feed on matter 'in the extremely well ordered state of more or less complicated organic compounds' rather than on charcoal or diamond pulp. He is right. . . . And so Simon is quite right in pointing out to me, as he did, that actually the energy content of our food does matter; so my mocking at the menu cards that indicate it was out of place. Energy is needed to replace not only the mechanical energy of our bodily exertions, but also the heat we continually give off to the environment. And that we give off heat is not accidental, but essential. For this is precisely the manner in which we dispose of the surplus entropy we continually produce in our physical life process."

    26. Ibid., 6.10.
    27. Derrida, "Force and Signification," 3.
    28. Ibid., 15.
    29. Ibid., 31.

30. Ibid., 22. To conclude the reference to *Force and Signification*, it is worth remarking that for Derrida the privilege of form over force represents the inaugural gesture of the metaphysics of presence, which encompasses structuralism as well as phenomenology. Cf. J. Derrida, *Force and Signification*, 33. "That modern structuralism has grown and developed within a more or less direct and avowed dependence upon phenomenology suffices to make it a tributary of the most purely traditional stream of Western philosophy, which, above and beyond its anti-Platonism, leads Husserl back to Plato. Now, one would seek in vain a concept in phenomenology which would permit the conceptualization of intensity or force. The conceptualization not only of direction but of power, not only the *in* but the *tension* of intentionality. All value is first constituted by a theoretical subject."

31. Jacob, *The Logic of Life*, 251.

32. Ibid., 252.

33. Derrida, *La vie la mort*, 6.11.

34. Jacob, *The Logic of Life*, 253. It is from this point, that is, from the contestation of the possibility of reducing or subordinating energy to information, that we can imagine a confrontation between Derrida's work and the most recent biological theories that are grounded on this possibility, that is, those theories based on the theory of self-organized system, including the theory of complexity. I believe that the missed emphasis on this crucial point constitutes the limit of the work of Christopher Johnson, who however has the merit of showing the interest and legitimacy of this confrontation, despite his inability to read Derrida's unedited texts. Cf. Johnson, *System and Writing*, 142–200. In fact, the formalization of the necessary openness of self-organized systems and the interest that these theories have in phenomena of such as noise, disorder, randomness, as positive factors of the transmission and integration of information and thus of the constitution of self-organized systems, offer an evident space of confrontation with Derrida's work. However, we must not forget that for Derrida the alterity that inhabits the living as its irreducible condition of possibility is of the order of the absolute alterity: it is not a merely aleatory disorder, that can be integrated in the system. For the living, absolute alterity cannot be but (the possibility of) death. It is from this irreducible relationship to the absolute alterity that we must rethink the conditions of possibility of the relations to alterity in its multiple manifestations and thus also to the aleatory that characterizes the relations that a system entertains with its internal and external environment. Here I do not mean to exclude a possible confrontation that, conversely, I consider necessary. I limit myself to laying out some coordinates in a field that is extended and differentiated and thus requires to be explored somewhere else. Cf. Atlan, *L'organisation biologique et la théorie de l'information*; Atlan, *Entre le cristal et la fumée: essai sur l'organisation du vivant*. Atlan, *Selected Writings: On Self-Organization, Philosophy, Bioethics, and Judaism*. On the theory of complexity, see also the work of Edgar Morin; to begin with, Morin, *Le paradigme perdu: la nature humaine*.

35. Derrida, *La vie la mort*, 6.12.
36. Derrida, "Freud and the Scene of Writing," 266.
37. N. Wiener, *The Human Use of the Human Being*, 95.
38. Jacob, *The Logic of Life*, 252.
39. Ibid., 271.
40. Derrida, *La vie la mort*, 6.13.
41. Jacob, *The Logic of Life*, 287.
42. Derrida, *La vie la mort*, 6.19.

Chapter V

1. It is worth remarking that in the *Standard Edition* of Freud, the word "*Trieb*," "drive," is systematically translated as "instinct." I shall quote Derrida's text from "To Speculate—On Freud," that is, from the version of the seminar published in *The Post Card*. When I refer to this text, I will also put between parentheses the reference to the corresponding passage in the seminar.
2. Derrida, "Freud and the Scene of Writing," 253.
3. Ibid., 254.
4. Derrida, *La vie la mort*, 11.1.
5. Freud, *Beyond the Pleasure Principle*, 7.
6. Ibid.
7. Ibid., 10.
8. Ibid.
9. Derrida, *The Post Card*, 282 (*La vie la mort*, 11.13).
10. Ibid., 283.
11. Ibid., 284 (*La vie la mort*, 11.14).
12. Cf. Derrida, "Living On."
13. Derrida, *The Post Card*, 285 (*La vie la mort*, 11.14). The English translation refers to *Living On* in the corresponding note, whereas in the original text (as well as in the seminar), the reference is between parentheses.
14. Derrida, *The Post Card*, 285 (*La vie la mort*, 11.14). Cf. J. Bennington, *Derridabase*, "The Unconscious," 138: "The pleasure-principle here names this setup in which the reality-principle serves it by putting obstacles in its way which oblige it to seek its goal via the detour of *différance*. Pure pleasure and pure reality would be equally mortal. Life is in their *différance*. It follows that the reality-principle is not in opposition to the pleasure-principle, but that it is the same thing, in *différance*, the detour via which the pleasure-principle rules and rules itself. But even this detour cannot be absolute (we know that *différance* cannot be absolute)—for it is nothing other than the passage of pleasure through the constraints of reality. The pleasure-principle is thus not other than the reality-principle, which it would

become absolutely if the detour did not finally return to pleasure. Pleasure is in the end nothing other than the passage of its own detour through reality, and it thus never arrives at its purity, which would again be death. We are still in a structure of the nonidentical same, which Derrida here calls 'life-death.'"

15. Derrida, *The Post Card*, 290 (*La vie la mort*, 11. 19).
16. Freud, *Beyond the Pleasure Principle*, 26.
17. Ibid.
18. See Helmoltz, "On the Thermodynamics of Chemical Processes."
19. Laplanche, *Life and Death in Psychoanalysis*, 119.
20. Freud, *Beyond the Pleasure Principle*, 27.
21. Ibid., 28.
22. Ibid., 29.
23. Ibid., 30.
24. Ibid.
25. Ibid., 12: "'Fright,' 'fear' and 'anxiety' are improperly used as synonymous expressions; they are in fact capable of clear distinction in their relation to danger. 'Anxiety' describes a particular state of expecting the danger or preparing for it, even though it may be an unknown one. 'Fear' requires a definite object of which to be afraid. 'Fright,' however, is the name we give to the state a person gets into when he has run into danger without being prepared for it; it emphasizes the factor of surprise. I do not believe anxiety can produce a traumatic neurosis. There is something about anxiety that protects its subject against fright and so against fright-neuroses."
26. Ibid., 32.
27. Derrida, *The Post Card*, 350 (*La vie la mort*, 13.7).
28. Freud, *Beyond the Pleasure Principle*, 34.
29. Ibid.
30. Derrida, *The Post Card*, 351 (*La vie la mort*, 13.9).
31. Freud, *Beyond the Pleasure Principle*, 35.
32. Ibid., 38.
33. Derrida, *The Post Card*, 354 (*La vie la mort*, 13.10).
34. It is evident here that Derrida is referring to Jacob's text, *The Logic of Life*.
35. Derrida, *The Post Card*, 355 (*La vie la mort*, 13.11).
36. Ibid., 394.
37. Derrida, *The Post Card*, 356.
38. Ibid., 356. Cf. G. Bennington, *Derridabase*, "The Unconscious," 141: "According to Freud, life (thus complicated) is a detour of the inorganic toward itself: the pleasure-principle defers the mortal cathexis or decathexis in the service of a movement, insured by the partial drives, toward a death which would be proper to the living being, the proper of the living being thus being to reappropriate to itself the very thing (death) which disappropriates it. Abyss of the proper.

The self of the living being is constituted as this detour toward its proper, its death."

39. Derrida, *The Post Card*, 359.

40. Cf. Freud, *Beyond the Pleasure Principle*, 42: "The repressed instinct [*Trieb*] never ceases to strive for complete satisfaction, which would consist in the repetition of a primary experience of satisfaction. No substitutive or reactive formations and no sublimations will suffice to remove the repressed instinct's [*triebende*] persisting tension; and it is the difference in the amount between the pleasure of satisfaction which is *demanded* and that which is actually achieved that provides the driving factor which will permit of no halting at any position attained, but, in the poet's words, '*ungebändigt immer vorwärts dringt*' ['Presses ever forward unsubdued'].* The backward path that leads to complete satisfaction is as a rule obstructed by the resistances which maintain the repression. So there is no alternative but to advance in the direction in which growth is still free—though with no prospect of bringing the process to a conclusion or of being able to reach the goal." *J. W. Goethe, *Faust*, I, 4.

41. Derrida, *The Post Card*, 362.

42. Cf. Freud, *Beyond the Pleasure Principle*, 45: "Perhaps we have adopted the belief because there is some comfort in it. if we are to die ourselves, and first to lose in death those who are dearest to us, it is easier to submit to a remorseless law of nature, to the sublime Necessity, than to a chance which might perhaps have been escaped. It may be, however, that this belief in the internal necessity of dying is only another of those illusions which we have created '*um die Schwere des Daseins zu ertragen*'* ['To bear the burden of existence'] . . . We must therefore turn to biology in order to test the validity of the belief." *F. Schiller, *Die Braut von Messina*, I, 8.

43. Derrida, *The Post Card*, 363 (*La vie la mort*, 13.13)

44. Entrambi fanno riferimento a *Über die dauer des Lebens* (1882), *Über Leben und Tod* (1884) e *Das Keimplasma: eine Theorie der Vererbung* (1892). Cf. Weismann, *Essays upon Heredity*; and Weismann, *The Germ-Plasm, A Theory of Heredity*.

45. Jacob, *The Logic of Life*, 217.

46. Freud, *Beyond the Pleasure Principle*, 46.

47. Ibid.

48. Ibid., 49.

49. Cf. Ibid., 49–50.

50. Ibid., 50.

51. Derrida, *The Post Card*, 365 (*La vie la mort*, 13.14).

52. Cf. Freud, *Beyond the Pleasure Principle*, 58: "Shall we follow the hint given us by the poet-philosopher, and venture upon the hypothesis that living substance at the time of its coming to life was torn apart into small particles, which have ever since endeavoured to reunite through the sexual instincts [*Sexualtriebe*]? That these instincts [*Triebe*], in which the chemical affinity of inanimate matter persisted,

gradually succeeded, as they developed through the kingdom of the protista, in overcoming the difficulties put in the way of that endeavour by an environment charged with dangerous stimuli—stimuli which compelled them to form a protective cortical layer? That these splintered fragments of living substance in this way attained a multicellular condition and finally transferred the instinct [*Trieb*] for reuniting, in the most highly concentrated form, to the germ-cells?—But here, I think, the moment has come for breaking off."

53. Ibid., 55.
54. Ibid., 62.
55. Ibid.
56. Derrida, *The Post Card*, 393 (*La vie la mort*, 14.4).
57. Ibid., 394 (*La vie la mort*, 14.4).
58. Ibid., 389.
59. Derrida, "Force and Signification," 22.
60. Derrida, *The Post Card*, 396 (*La vie la mort*, 14.6).
61. Freud, *Beyond the Pleasure Principle*, 62.
62. Derrida, *The Post Card*, 396 (*La vie la mort*, 14.6)
63. Cf. Freud, *Beyond the Pleasure Principle*, 63.
64. Derrida, *The Post Card*, 400 (*La vie la mort*, 14.7). Cf. G. Bennington, *Derridabase*, "The Unconscious," 141: "The pleasure-principle binds the freely circulating energy of the primary processes. For there to be pleasure, the pleasure-principle must limit pleasure, which would otherwise be absolute unpleasure and short-circuit in the burnout of an im-proper death. Pleasure begins by binding *itself* or limiting *itself* in order to be what it is. There is no (absolute) pleasure, but by the same token there is no (absolute) unpleasure. This band and contraband, this stricture of the pleasure-principle constitutes reality as the very tension of self-binding pleasure. No pleasure without stricture. There is no lack or opposition in this logic, desire is here, "productive," certainly, but only in limiting its "production"—we cannot say that the more it binds, the more pleasure there is, nor the opposite, we're always dealing with more and less."
65. Derrida, *The Post Card*, 402.
66. Ibid (*La vie la mort*, 14.9).
67. Cf. Freud, *Beyond the Pleasure Principle*, 16.
68. Derrida, *The Post Card*, 403 (*La vie la mort*, 14.9).
69. Freud, *Beyond the Pleasure Principle*, 54.
70. Derrida, *The Post Card*, 404.
71. Ibid., 405.
72. Ibid. Cf. G. Bennington, *Derridabase*, "The Unconscious," 143: "This is also the place to talk about mastery. The whole discussion of the pleasure-principle turns around its mastery in the negotiation between primary processes and reality. Freud also talks, in passing, of a drive to mastery, or, as Derrida translates it into French, of *emprise*. 'Quasi-transcendental' privilege of this drive: one drive in the series of the

drives, it also says the being-drive of drives, the driveness of the drive. Every drive must retain a relation to itself (as other) which binds it to itself, if it is to be the drive it is—this is how deconstruction formulates the law of identity in general (PC, 403). And psychic life in general is described as a power game between drives, and between the drives and the pleasure-principle (PP, the supposed master). (In French, 'PP' is pronounced 'pepe,' slang for 'grandfather,' and Derrida plays extensively on this in *The Post Card*). This question of mastery thus logically precedes the question of pleasure and unpleasure: *différance* of power, of forces still. . . . Mastery/emprise speaks of a relation to the other, who can also be oneself. This is one of the most constant themes: in order to be itself, a subject must already refer to itself as to an other. Identity comes only from alterity, called by the other."

73. Cf. Derrida, *Audelà du principe de pouvoir*.
74. Derrida, "Psychoanalysis Searches the States of Its Soul," 258.
75. Ibid., 271. Here I note that, in the same text, Derrida highlights Freud's interest in the life sciences and above all in genetics. He justifies this interest at a methodological level by referring to the evidently heuristic function of the general text as the differential/differing condition of the living and its products. Cf. ibid., 244: "As for the physical, neuronal, or genetic sciences, Freud was the first not to reject, but to expect a lot from them—provided that one knows how to wait expectantly, precisely, and to articulate without confusing, without precipitously homogenizing, without crushing the different agencies, structures, and laws, while respecting the relays, the delays, and, do I dare say, the deferred of differance."
76. Ibid., 241.
77. Ibid., 268.
78. Derrida, *The Beast and Sovereignty I*, 388.

Chapter VI

1. Derrida, *Faith and Knowledge*, 73n.
2. This is the case, I believe, of Hägglund's interpretation: in *Radical Atheism* he aims to read Derrida's recourse to the autoimmunitarian beyond the reference to autoimmunity: cf. Hägglund, *Radical Atheism*, 9: "I develop the logic of autoimmunity throughout this book, but I want to point out that I am not concerned with the relation between how Derrida uses the term 'autoimmunity' and how it is employed in biological science. Autoimmunity is for me the name of a deconstructive logic that should be measured against the standards of philosophical logic. This does not mean that the biological connotations of the term are not important, but they do not make the argument dependent on its correspondence with discoveries in contemporary science. The biological connotations of the term 'autoimmunity' remind us that Derrida pursues a logic of life (or, rather, life-death), but I seek to establish the power of this logic on philosophical rather than scientific grounds."

However, the resistance to engaging with biology leads Hägglund to interpret the autoimmunitarian as a negative consequence of the unconditioned opening of the (living) identity to alterity, namely, of the finitude of the living, and not as the structural condition of this opening, and thus to privilege the example of autoimmune pathologies in order to explain Derrida's recourse to the autoimmune: cf. ibid., 15: "Thus, if there is no indivisible identity, every immune system runs the risk of being autoimmune, since there can be no guarantee that it will be in the service of maintaining health. What is attacked as an enemy of the body may turn out to be an essential part of the body, and what is welcomed as beneficial to the body may turn out to destroy the body from within." From the perspective of biology, an immunitarian system unable to distinguish between pathogen and positive agents is affected by a serious autoimmune pathology, for whom, according to Hägglund, there is no cure: ibid., 48: "The defense of life is thus attacked from within. There can be no cure for such autoimmunity since life is *essentially* mortal."

3. Derrida, *Faith and Knowledge*, 70n.
4. Derrida, "Above All No Journalists!," 67.
5. Derrida, *The Animal That Therefore I Am*, 47.
6. Derrida, *Rogues*, 55.
7. Ibid., 123.
8. Ibid., 109. The absence of limits in autoimmunitarian logic is reaffirmed in "The 'World' of the Enlightenment to Come," the second essay of *Rogues*, where Derrida analyzes through Husserl the crisis and rescue of reason. Cf. ibid., 124: "As I have done elsewhere, I have here granted to this autoimmune schema a range without limit, one that goes far beyond the circumscribed biological processes by which an organism tends to destroy, in a quasi-spontaneous and more than suicidal fashion, some organ or other, one or another of its own immunitary protections." Cf. also Derrida, "Autoimmunity: Real and Symbolic Suicides," 94, where Derrida designates the autoimmunitarian logic as "an implacable law" and as "without limits."
9. See also Derrida, "Autoimmunity: Real and Symbolic Suicides," 187n: "For example, in *Faith and Knowledge: The Two Sources of 'Religion' at the Limits of Reason Alone*, . . . In analyzing 'this terrifying but inescapable logic of the autoimmunity of the unscathed that will always associate Science and Religion,' I there proposed to extend to life in general the figure of an autoimmunity whose meaning or origin first seemed to be limited to so-called natural life or to life pure and simple, to what is believed to be the purely 'zoological,' 'biological,' or 'genetic.'"
10. Derrida, *Rogues*, 45.
11. Cf. Derrida, *Specters of Marx*, 67, 185, 235. I note that here Derrida links the question of life-death to the possibility of thinking "survival," which I discuss in the next chapter: "And this question would be a question of life or death, the question of life-death, before being a question of Being, of essence, or of existence. It would open onto a dimension of irreducible survival or surviving [survivance] and onto Being and onto some opposition between living and dying."

12. Derrida, *Specters of Marx*, 177.
13. Derrida, *Faith and Knowledge*, 73n.
14. The most recent literature on this topic speaks about "natural" or "physiological," autoimmunity, and "auto-antibodies" (Auto-Ab) as its agents. For an overview of this literature, see A. B. Poletaev, L. P. Churilov, Yu. I. Stroev, and M. M. Agapov, "Immunophysiology versus Immunopathology: Natural Autoimmunity in Human Health and Disease," 2012. For the authors of this essay, immunology acknowledged the phenomena of natural autoimmunity only at a later stage, because of their "microbiological matrix." By virtue of this matrix, the immunitary system had been interpreted, for a long time, exclusively as a defense system, according to that strategic and military model Roberto Esposito critically refers to in *Immunitas*, a book that, as we will see, only takes into account the immunology of microbiological matrix, thus forgetting the phenomena of natural auto-immunity. I quote from the aforementioned essay: "Historically, immunology emerged as a branch of applied microbiology. Therefore 'microbiological' thinking, namely its idea of war against aliens, has persisted in minds for decades due to the fact that generations of immunologists have been educated by microbiologists. The cells of the immune system were metaphorically interpreted as 'gendarmes' or 'border guards'; first this allegory was probably coined in 1847 by Virchow and brightly expressed much later (1896) by Duclaux" (Poletaev et al., "Immunophysiology versus Immunopathology," 221). The physiological and pathophysiological approach allowed us to recognize the essential role that natural autoimmunity plays in the construction of the organism since the process of cell differentiation across the embryonal development and, more generally, in the conservation of the life of the living: "In pathophysiological approach, the systems supervising the growth, development and aging of the whole organism and all its components are supposed to coordinate primarily the sequence and intensity of reading of the genetic information in various cells. This task can be achieved neither by neural mechanisms, nor via hormonal agents, which are involved in accelerating or slowing down the metabolic processes. The neurotransmitters, hormones and their receptors lack ontogenetic and event-driven variability needed for this purpose, while the cells producing them lack the necessary mobility and all-embracing dispersal—the qualities *inherent to the immune system*. Transfer of the emphasis in the main purpose of the immune system from defense to homeodynamic regulation will necessarily lead to reevaluation of some conventional views, for example, such as the phenomena of physiological autoimmunity and the general role of natural autoantibodies (auto-Abs). In our opinion, it is physiological autoimmunity that provides for bringing together and co-tuning of genetic information processing in different cells of the holistic 'Body' through the complete ontogeny" (ibid., 223). Hence, it follows that, according to the authors of this volume, the immunitary system of an accomplished organism is a secondary and adaptive structure with respect to the primary process of construction based on an autoimmunitary dynamic. Therefore, to avoid confusions between natural

and pathological autoimmunity, between autoimmunity—see the case of Derrida's autoimmunity—and autoimmune diseases, a more sophisticated terminological distinction would be required: "How can sanogenic (beneficial) autoimmune reactions be differentiated from the *pathogenic* (harmful) ones? It seems that primary autoimmune reactions (the ones not justified by real needs of an organism) in most cases are poorly regulated. Immune response may sometimes be inadequate in intensity, or incorrectly targeted, or badly driven, and in each case the result may be rather detrimental for an organism. All these situations, when immune response brings more harm than defense, are referred in Pathophysiology by the collection term '*allergy*.' From a didactical point of view, we propose to use the term '*autoimmunity*' preferably in relation to physiological autoimmune processes. Adaptive secondary autoreactive responses (autoimmune) should be distinguished from '*autoallergic*' (mostly, primary) pathogenic immune reactions. We recommend using the term '*autoallergy*' for abnormal rise in production of auto-Ab(s) and autoreactive lymphocytes (provoked by viruses, bacteria, chemicals or other harmful factors, or related to disorders in the regulation of natural autoimmunity and non-conditioned by real needs of an organism)" (ibid., 229).

15. I recall that Michael Naas finds in the reference to immunodepressants the positive bearing of autoimmunitarian bio-logic, cf. Naas, *Derrida from Now On*, 131: "in *Faith and Knowledge* autoimmunity is presented not only as a threat but as a chance for any living organism: a threat insofar as it compromises the immune system that protects the organism from external aggression, but as in the case of immuno-depressants, a chance for an organism to open itself up to and accept something that is not properly its own, the transplanted organ, the graft, in a word, the other, which is but the cutting edge, the living edge, of the self. Without certain forces of autoimmunity, we would reject organs and others essential to 'our' survival." See also Naas, *Miracle and Machine*, 82. My purpose is to highlight the biological nature of these forces.

16. Derrida, *Faith and Knowledge*, 51.

17. Derrida, "Autoimmunity: Real and Symbolic Suicides," 145.

18. Cf. Ameisen, *La sculpture du vivant*, 37, 40, 66, 132, 147, 190, 200, 284, 316, 335, 422, 438. The biological paradigm adopted by Ameisen, who was a student of Henri Atlan, is still scriptural, cf. ibid., 30: "the extraordinary diversity of forms, activities, functions and potentialities of the cells that compose our body is determined by the interactions among the tools, that is, the proteins, that our cells ceaselessly produce on the basis of the information contained in the genes. The genes are like large sequences of letters that are organized in words and connected in sentences and form books. The library of our genes is constituted by thirty to forty thousand books, all different and each present in a double specimen, the one coming from the father and the other from the mother, two variations on the same theme, like two different editions, marked and revised, of the same book. The language in which the books of our genes are written is common to the

whole universe of the living, from bacteria to birds, from flowers to fish, from rats to man. The only thing that changes, in each species, is the nature of the text. The vocabulary of this universal language includes sixty-four words, each formed by a sequence of three letters, on the basis of an alphabet of extreme simplicity that consists of four letters. This alphabet has a concrete nature: each letter is a molecule. The library of the books of our genes thus unfolds itself through a chain of several billions of letters that constitute the immense filament of DNA." This is to recall that, "at least" in biology, our age has not decided yet to abandon the heuristic value of writing. At this point, we cannot but recognize in this matter of fact an ineludible necessity for the sciences *of* the living (which, of course, should be read as a subjective as well as objective genitive).

19. Ibid., 38.
20. Ibid., 41
21. Ibid., 46.
22. Ibid., 61.
23. Ibid., 69.
24. Ibid., 76.
25. Ibid., 74.
26. Ibid., 76.
27. Ibid., 82.
28. Ibid., 216.
29. Ibid., 221.
30. Ibid., 102.
31. Ibid., 104.
32. Ibid. 135.
33. On this use of "phantasm," see Naas, *Miracle and Machine*, 209: "This indemnification would be related to a ritual sacrifice that protects or compensates, that reconstitutes or attempts to reconstitute some intact purity, and that does so, oftentimes, upon the body itself, by marking or doing violence to the body itself, or else by replacing it with a kind of phantasm, namely, the phantasm of a body with total immunity. Though such a body might then seem to be most living, most protected from death, it would be in the end, as a body closed up within itself, closed off not only from what can harm it but from what can save it or allow it to live on, a body at once completely alive and completely dead."
34. Derrida, *Rogues*, 152.
35. Ibid., 123.
36. Derrida, *Faith and Knowledge*, 51. According to Roberto Esposito, immunology, the discourse about the description and definition of the immunitary system, constitutes the very paradigm of biopolitics in its contemporary declination. cf. Esposito, *Immunitas*, 150: "Taking this semantic crossroads as our point of departure, the immune system is revealed as the nerve center through which the political governance of life runs. The immune system pushes the governance of life beyond

the biopolitical paradigm or gives another meaning to the paradigm of biopolitics, one that is different from its usual formulation. What drops away is precisely the presumption of a direct and immediate relationship between politics and life. In reality, they are increasingly related through the great figurative device that medical science has developed around the body's need for self-protection. From this, from this concentration of real and metaphoric functions, a semantic wave destined to wash over the entire gamut of social languages is set in motion. No wonder; if the semiotic axis around which every social institution is constituted lies in the boundary between self and other—between us and them—what constitutes both its interpretative key and effective outcome better than the principle of immunity?" Within this order of the immunological discourse, Esposito also inscribes the theory of cellular suicide, namely, apoptosis, thus reducing it to the general scheme of the defense of the immunity of the body proper against the other, the stranger (ibid., 159). However, he limits himself to merely mentioning that notion without dedicating to it an explicit exposition. Nor does he refer to Ameisen. Therefore, he does not recognize the critical and deconstructive significance of that notion with respect to the hegemonic, immunological discourse, which is governed by the exclusive opposition between identity and otherness. In particular, Esposito does not see that apoptosis does not only concern the functioning of the immunitary system but also the construction of the organism since its embryonal stage and thus of those organs that have an immunitary privilege. Esposito himself, in fact, seems to attribute to that phenomenon the possibility of overcoming the immunological paradigm, in view of a relation or exposure to the other that is no longer understood as a mere threat to be conjured away. It is precisely from this perspective that apoptosis could have been fertile also for Esposito, so long as it allows us to conceive of death—a figure at the same time as a matrix of absolute alterity—as the internal and irreducible condition of the genesis and structure of the living and thus to conceive of its effects on the construction and functioning of the immunitary system. Here I refer to the necessity of the auto-immunitary process in which the organism destroys its own defenses as a condition for its relationship with the other, which is necessary to the life of the living. In fact, Esposito does not even take into account the better-known phenomena of "natural autoimmunity"; rather, he interprets autoimmunity as an exclusively pathological phenomenon. One could suppose that this is just an omission from the large panorama of the scientific literature on this topic (a risk from which this book too cannot consider itself immune). However, one cannot forget that the theory of cellular suicide makes tremble the supposed homogeneity of the immunological discourse and thus its paradigmatic value. What is at stake here is remarkable: for Esposito, the immunitary paradigm constitutes a problem, as it is grounded on the classical and metaphysical ("mythical") notion of identity (which would be self-constituted, autonomous, unconditioned, and thus independent from the other and exclusive) and thus is unable to account for the vulnerability of the human body and of its irreducible finitude (ibid., 159). In order to overcome this paradigm and thus conjure the biopolitical tradition away, it would be required, or merely sufficient, admitting that—this is Esposito's conclusion—"before any other transformation, each body is already exposed to the need for its own exposure. This is the condition common to all that is immune: the endless perception of its own finitude" (ibid., 174). This emphasis on our finitude, on the vulnerability of a body that constitutes itself in the relationship with the other to which it must be fatally exposed, testifies that for Esposito too death is merely the other of life, its external limit, to which we are thus exposed because of our ontological finitude and not an irreducible and internal condition for the life of the living, for its genesis as well as structure, before any contingent exposure to the otherness of the other (whether a virus or a stranger) and as an irreducible condition for the exposure itself, according to the illogic logic of life-death elaborated by Derrida. These critical remarks, developed in the space of a footnote, do not resolve the debate between biopolitics and (bio) deconstruction that I consider necessary and full of repercussions on both sides, in particular, apropos of its possible declinations within a political horizon, a horizon that goes beyond the methodological limits of this book.

Chapter VII

1. Derrida, "Final Words," 462.
2. Derrida, *Learning to Live Finally*, 96.
3. Cf. Derrida, "A Number of Yes," 232.
4. Derrida, "A 'Madness' Must Watch Over Thinking," 96.
5. Derrida had already posited this articulation in 1964 when he used for the first time the notion of survival in relation to the dimension of affirmation: cf. Derrida, "Edmond Jabés and the Question of the Book," 95: "Life negates itself in literature only so that it may survive better. So that it may be better. It does not negate itself any more than it affirms itself: it differs from itself, defers itself, and writes itself as différance."
6. In what follows I refer to the first, English version of "Living On" in H. Bloom et al., *Deconstruction and Criticism*.
7. Derrida, "A Number of Yes," 232.
8. Ibid., 236.
9. Ibid., 238.
10. Ibid., 239.
11. Ibid., 238–239.
12. Derrida, "Living On," 87.
13. Here I refer to Derrida's early works on Husserl and to *Of Grammatology* and, in particular, to those texts in which the deconstruction of Husserl's living present is taken up through the notions of "arche-trace" and "arche-writing" and

is extended to the metaphysics of presence in general. On speech act theory, see, for instance, Derrida, "Signature, Event, Context' in Derrida," 307–330. In particular, 322: "Austin has not taken into account that which in the structure of locution (and therefore before any illocutory or perlocutory determination) already bears within itself the system of predicates that I call *graphematic in general*, which therefore confuses all the ulterior oppositions whose pertinence, purity, and rigor Austin sought to establish in vain. In order to show this, I must take as known and granted that Austin's analyses permanently demand a value of context, and even of an exhaustively determinable context, whether de jure or teleologically; and the long list of 'infelicities' of variable type which might affect the event of the performative always returns to an element of what Austin calls the total *context*. One of these essential elements—and not one among others—classically remains consciousness, the conscious presence of the intention of the speaking subject for the totality of his locutory act. Thereby, performative communication once more becomes the communication of an intentional meaning, even if this meaning has no referent in the form of a prior or exterior thing or state of things. This conscious presence of the speakers or receivers who participate in the effecting of a performative, their conscious and intentional presence in the totality of the operation, implies teleologically that no remainder escapes the present totalization. No *remainder*, whether in the definition of the requisite conventions, or the internal and linguistic context, or the grammatical form or semantic determination of the words used; no irreducible polysemia, that is no 'dissemination' escaping the horizon of the unity of meaning."

14. Derrida, "Living On," 103.
15. Ibid., 109.
16. Ibid., 112.
17. Ibid., 114.
18. Ibid., 132.
19. Ibid.
20. Ibid., 136.
21. Ibid., 138.
22. Ibid., 136.
23. Ibid., 102.
24. Allow me to refer to my reading of the conceptual web experience-trace-testimony in Vitale, "Let the Witness Speak: From Arche-writing to the Community To Come," and on testimony and survival in Vitale, "Conjuring Time: Jacques Derrida Between Testimony and Literature." It was by working on "Demeure" that the urgency of engaging with the "early" Derrida around the question of the survival of the living (on) emerged.
25. Derrida, "Freud and the Scene of Writing," 254.
26. Ibid., 282.
27. Jacob, *Logic of Life*, 1.

28. Ibid., 254.
29. Derrida, *La vie la mort*, 5.1.
30. Derrida, *La vie la mort*, 6.2.

Bibliography

Ameisen, Jean Claude. "Programmed Cell Death and AIDS: From Hypothesis to Experiment." *Immunology Today* 13, no. 10 (1992): 388–391.
———. *La sculpture du vivant: Le suicide cellulaire ou la mort créatrice*. Paris: Seuil, 1999.
Ameisen, Jean Claude, and André Capron. "Cell Dysfunction and Depletion in AIDS: The Programmed Cell Death Hypothesis." *Immunology Today* 12, no. 4 (1991): 102–105.
Aristotle. *On the Parts of Animals*. Translated by James G. Lennox. Oxford: Oxford University Press, 2001.
Atlan, Henry. *L'organisation biologique et la théorie de l'information*. Paris: Hermann, 1972.
———. *Entre le cristal et la fumée: essai sur l'organisation du vivant*. Paris: Seuil, 1979.
———. *Selected Writings: On Self-Organization, Philosophy, Bioethics, and Judaism*. Edited by Stefanos Geroulanos and Todd Meyers. New York: Fordham University Press, 2011.
Bennington, Geoffrey. *Derridabase*. In Geoffrey Bennington and Jacques Derrida, *Derrida*, translated by Geoffrey Bennington. Chicago: University of Chicago Press, 1993.
Blanchot, Maurice. *Death Sentence*. Translated by Lidia Davis. Barrytown, NY: Station Hill Press, 1998.
———. *The Folly of the Day*. Translated by Lidia Davis. Barrytown, NY: Station Hill Press, 1981.
———. "The Instant of my Death." In Maurice Blanchot and Jacques Derrida, *The Instant of My Death/Demeure: Fiction and Testimony*, translated by Elizabeth Rottenberg. Stanford: Stanford University Press, 2005.
Canguilhem, George. "The Concept of Life." In *A Vital Rationalist. Selected Writings from George Canguilhem*, edited by François Delaporte. New York: Zone Books, 1994.
Debru, Claude, Michel Morange, and Frédéric Worms. *Une nouvelle connaissance du vivant: François Jacob, André Lwoff, Jacques Monod*. Paris: Rue D'Ulm, 2012.

Derrida, Jacques, and Elisabeth Rudinesco. *For What Tomorrow . . . A Dialogue*. Translated by Jeff Fort. Stanford: Stanford University Press, 2004.

Derrida, Jacques. "Above All No Journalists!" Translated by Samuel Weber. In *Religion and Media*, edited by Hent de Vries and Samuel Weber. Stanford: Stanford University Press, 2001.

———. "A 'Madness' Must Watch Over Thinking." In *Points . . . Interviews 1974–1994*, edited by Elisabeth Weber. Stanford: Stanford University Press, 1995.

———. *The Animal That Therefore I Am*. Translated by David Wills. New York: Fordham University Press, 2008.

———. "A Number of Yes." In *Psyché: Inventions of the Other*. Vol. II. Stanford: Stanford University Press, 2008.

———. "Au-delà du principe de pouvoir." *Rue Descartes* 3, no. 82 (2014): 4–13.

———. "Autoimmunity: Real and Symbolic Suicides: A Dialogue with Jacques Derrida." In Giovanna Borradori, *Philosophy in a Time of Terror: Dialogues with Jürgen Habermas and Jacques Derrida*. Chicago: University of Chicago Press, 2003.

———. "Demeure: Fiction and Testimony." In Maurice Blanchot and Jacques Derrida, *The Instant of My Death/Demeure: Fiction and Testimony*, translated by Elizabeth Rottenberg. Stanford: Stanford University Press, 2005.

———. "Différance." In *Margins of Philosophy*, translated by Alan Baas. Chicago: University of Chicago Press, 1982.

———. *Dissemination*. Translated by Barbara Johnson. London: The Athlone Press, 1981.

———. "Edmond Jabés and the Question of the Book." In *Writing and Difference*, translated by Alan Baas. London: Routledge, 2002.

———. *Edmund Husserl's Origin of Geometry: An Introduction*. Translated by John P. Leavey Jr. Lincoln: University of Nebraska Press, 1989.

———. *Faith and Knowledge: The Two Sources of "Religion" at the Limits of Reason Alone*. In *Religion*, translated by S. Weber and edited by Jacques Derrida and Gianni Vattimo. Cambridge: Polity Press, 1998.

———. "Final Words." Translated by Gila Walker. *Critical Enquiry* 33, no. 2 (Winter 2007): 462.

———. *Force and Signification* in *Writing and Difference*. Translated by Alan Baas. London: Routledge, 2002.

———. "Freud and the Scene of Writing." In *Writing and Difference*, translated by Alan Baas. London: Routledge, 2002.

———. "From Restricted to General Economy." In *Writing and Difference*, translated by Alan Baas. London: Routledge, 2002.

———. *Glas*. Translated by John P. Leavey Jr. and Richard Rand. Lincoln: University of Nebraska Press, 1986.

———. "Interpreting Signatures (Nietzsche/Heidegger): Two Questions." In *Dialogue and Deconstruction: The Gadamer-Derrida Encounter*, edited by Diane P.

Michelfelder and Richard E. Palmer. Albany: State University of New York Press, 1986.

———. *La vie la mort*. Unpublished seminar, Archive-Derrida, IMEC, DRR 173.

———. *Learning to Live Finally. The Last Interview*. Translated by Pascale-Anne Brault and Michael Naas. New York: Palgrave Macmillan, 2010.

———. *Limited Inc*. Translated by Samuel Weber. Evanston: Northwestern University Press, 1988.

———. "Living On." In *Deconstruction and Criticism*, edited by Harold Bloom, Paul De Man, Jacques Derrida, Jeoffrey H. Hartmann, and John Hillis Miller. London: Routledge and Kegan Paul, 1979.

———. *Of Grammatology*. Translated by Gayatri Chakravorty Spivak. Baltimore: John Hopkins University Press, 1997.

———. *Otobiographies*. In *The Ear of the Other*, translated by Avital Ronell. New York: Schocken Books, 1985.

———. *Parages*. Translated by John Leavey and Tom Conley. Stanford: Stanford University Press, 2011.

———. "Poetics and Politics of Witnessing." In *Sovereignties in Question: The Poetics of Paul Celan*, edited by Thomas Dutoit and Outi Pasanen. New York: Fordham University Press, 2005.

———. *Positions*. Translated by Alan Baas. Chicago: University of Chicago Press, 1981.

———. *The Post Card: From Socrate to Freud and Beyond*. Translated by Alan Baas. Chicago: University of Chicago Press, 1987.

———. "Psychoanalysis Searches the States of Its Soul." In *Without Alibi*, translated by Peggy Kamuf. Stanford: Stanford University Press, 2002.

———. *Rogues. Two Essais on Reason*. Translated by Pascale-Anne Brault and Michael Naas. Stanford: Stanford University Press, 2005.

———. "Signature, Event, Context." In *Margins of Philosophy*, translated by Alan Baas. Chicago: Chicago University Press, 1982.

———. *Specters of Marx: The State of the Debt, the Work of Mourning and the New International*. Translated by Peggy Kamuf. New York: Routledge, 1994.

———. *Speech and Phenomena*. Translated by David B. Allison. Evanston: Northwestern University Press, 1973.

Diderot, Denis, and Jean Le Rond d'Alembert. *Encyclopédie, ou Dictionnaire raisonné des sciences, des arts et des métiers*. Paris: André le Breton, Michel-Antoine David, Laurent Durand, and Antoine-Claude Briasson: Paris, 1757.

Doyle, Richard. *On Beyond Living: Rhetorical Transformations of the Life Sciences*. Stanford: Stanford University Press, 1997.

Esposito, Roberto. "Comunità, immunità, biopolitica." In *Spettri di Derrida*, edited by Carola Barbero, Simone Regazzoni, and Amelia Valtolina. Genova: Il nuovo melangolo, 2010.

———. *Immunitas: The Protection and Negation of Life*. Translated by Zakiya Hanafi. Cambridge: Polity, 2011.

Favareau, Donald, ed. *Essential Readings in Biosemiotics: Anthology and Commentary*. Dordrecht: Springer, 2008.

Foucault, Michel. "Croître et multiplier." In *Dites et écrits. Tome II. 1970–1975*. Paris: Gallimard, 1994.

Francis, Richard C. *Epigenetics: the Ultimate Mystery of Inheritance*. New York: W. W. Norton & Company, 2011.

Freud, Sigmund. *Beyond the Pleasure Principle*. In *The Standard Edition of the Complete Psychological Works of Sigmund Freud*. Vol. 18, edited by James Strachey. London: Hogart Press and the Institute of Psychoanalysis, 1955.

———. *The Interpretation of Dreams*. In *The Standard Edition of the Complete Psychological Works of Sigmund Freud*. Vols. 4–5, edited by James Strachey. London: Hogart Press and the Institute of Psychoanalysis, 1953.

———. *Papers on Metapsychology*, In *The Standard Edition of the Complete Psychological Works of Sigmund Freud*. Vol. 14, edited by James Strachey. London: Hogart Press and the Institute of Psychoanalysis, 1957.

———. *A Project for a Scientific Psychology*. In *The Standard Edition of the Complete Psychological Works of Sigmund Freud*. Vol. 1, edited by James Strachey. London: Hogart Press and the Institute of Psychoanalysis, 1954.

———. *Three Essays on the Theory of Sexuality*. Vol. 7, edited by James Strachey. London: Hogart Press and the Institute of Psychoanalysis, 1962.

Gasché, Rodolphe. *The Tain of the Mirror: Derrida and the Philosophy of Reflection*. Cambridge: Harvard University Press, 1986.

Goethe, Johann Wolfgang von. *The Metamorphosis of the Plants*. Translated by Douglas Miller. Cambridge: MIT Press, 2009.

Hägglund, Martin. *Radical Atheism: Derrida and the Time of Life*. Stanford: Stanford University Press, 2008.

Hartsoeker, Nicolaas. *Essay de dioptrique*. Paris: Imprimerie Royale, 1694.

Hegel, Georg Wilhelm Friedrich. *Lectures on the Philosophy of Religion*. Vol. III, translated by E. B. Speirs and J. Burdon Sanderson. New York: The Humanities Press, 1962.

———. *The Philosophy of History*. Translated by John Sebree. Kitchener: Batoche Books, 1991.

———. *Philosophy of Nature*. Vol. III, edited by Michael John Petry. London: George Allen and Unwin, 1970.

———. *Science of Logic*. Translated by George Di Giovanni. Cambridge: Cambridge University Press, 2010.

———. "The Spirit of Christianity and Its Fate." In *On Christianity: Early Theological Writings*, translated by Thomas Malcolm Knox. New York: Harper and Brothers, 1971.

Helmoltz, Hermann. "On the Thermodynamics of Chemical Processes." In *Physical Memoirs Selected and Translated from Foreign Sources*. Vol. 1. London: Taylor and Francis, 1888.

Jablonka Eva, and Marion Lam. *Evolution in Four Dimensions: Genetic, Epigenetic, Behavioral, and Symbolic Variation in the History of Life*. Cambridge: MIT Press, 2005

Jacob, François. *The Logic of Life: A History of Heredity*. Translated by Betty E. Spillmann. New York: Pantheon Books, 1973.

Jämsä, Tuomo. "Semiosis in Evolution." In *Introduction to Biosemiotics: The New Biological Synthesis*, edited by Marcello Barbieri. Dordrecht: Springer, 2008.

Johnson, Christopher. *System and Writing in the Philosophy of Jacques Derrida*. Cambridge: Cambridge University Press, 1993.

———. "'French' Cybernetics." *French Studies* 69, no. 1 (2015): 60–78.

Kay, Lily E. "A Book of Life? How the Genoma Became an Information System and DNA a Language." *Perspectives in Biology and Medicine* 41, no. 4 (1998): 504–528.

———. *Who Wrote the Book of Life?: A History of the Genetic Code*. Stanford: Stanford University Press, 2000.

Kojève, Alexandre. *Introduction à la lecture de Hegel*. Paris: Gallimard, 1947.

Kull, Kalevi. "A sign is not alive—a text Is." *Sign System Studies* 30, no. 1 (2002): 327–336.

Laplanche, Jean. *Life and Death in Psychoanalysis*. Translated By Jeffrey Mehlman. Baltimore: John Hopkins University Press, 1976.

Leroi–Gourhan, André. *Gesture and Speech*. Translated by Anna Bostock Berger. Cambridge: MIT Press, 1993.

Lwoff, André. "Le concept d'information dans la biologie moléculaire." In "Cahiers de Royaumont," Philosophie N° V, *Le concept d'information dans la science contemporaine*. Paris: Minuit, 1965.

Malabou, Catherine. *Plasticity at the Dusk of Writing: Dialectic, Destruction, Deconstruction*. Translated by Carolyn Shread. New York: Columbia University Press, 2010.

———. "The End of Writing? Grammatology and Plasticity." *The European Legacy* 12, no. 4 (2007): 431–441.

Monod, Jacques. *Chance and Necessity: An Essay on the Natural Philosophy of Modern Biology*. Translated by Austryn Wainhouse. New York: Alfred A. Knopf, 1971.

Morange, Michel. "Introduction." In *Travaux scientifiques de François Jacob*, edited by Nadine Peyrieras and Michel Morange. Paris: Odile Jacob, 2002.

Morin, Edgar. *Le paradigme perdu. La nature humaine*. Paris: Seuil, 1973.

Naas, Michael. *Miracle and Machine: Jacques Derrida and the Two Sources of Religion, Science, and the Media*. New York: Fordham University Press, 2012.

———. *Derrida From Now On*. New York: Fordham University Press, 2008.

Narra, Hema Prasad, and Howard Ochman. "Of What Use Is Sex to Bacteria?" *Current Biology* 16, no. 17 (2006): R705–R710.

Nielsen, Kaare M., and Christopher M. Thomas. "Mechanisms of, and Barriers to, Horizontal Gene Transfer between Bacteria." *Nature Reviews Microbiology* 3, no. 9 (2005): 711–721.

Nietzsche, Friedrich. *On the Future of our Educational Institution*. In *The Complete Works of Friedrich Nietzsche*. Vol. 3, translated by John McFarland M. Kennedy. New York: Russel, 1964.

Peeters, Benoît. *Derrida: A Biography*. Translated by A. Brown. Cambridge: Polity Press, 2013.

Poletaev, Alexander B., Eugene Agapov, Leonid P. Churilov, and Yuri I. Stroev. "Immunophysiology versus Immunopathology: Natural Autoimmunity in Human Health and Disease." *Pathophysiology* 19, no. 3 (2012): 221–231.

Schrödinger, Erwin. *What Is Life?: The Physical Aspect of the Living Cell*. Cambridge: Cambridge University Press, 1992.

Sebeok, Thomas A. "Biosemiotics: Its Roots, Proliferation and Prospects." *Semiotica* 2001, no. 134 (2001): 61–78.

Senatore, Mauro. "Of Seminal Difference: Dissemination and Philosophy of Nature." *New Centennial Review* 15, no. 1 (2015): 67–92.

Stiegler, Bernard. *Technics and Time, 1. The Fault of Epimetheus*. Translated by Richard Beardsworth and George Collins. Stanford: Stanford University Press, 1998.

Swammerdam, Jan. *Miraculum naturae sive uteri muliebris factorya*. Amsterdam: van Horne, 1672.

von Uexkull, Jakob. *Umwelt und Innenwelt der Tiere*. Berlin: Julius Springer Verlag, 1909.

Vitale, Francesco. "Archiscrittura." In *Derridario. Dizionario della decostruzione*, edited by Silvano Facioni, Simone Regazzoni, and Francesco Vitale. Genova: Il nuovo Melangolo, 2012.

———. "Conjuring Time: Jacques Derrida between Testimony and Literature." *Parallax* 17, no. 1 (2011): 54–54.

———. "Let the Witness Speak: From Archi-writing to the Community to Come." *Derrida Today* 2, no. 2 (2009): 260–270.

———. *Mitografie. Jacques Derrida e la scrittura dello spazio*. Milano: Mimesis, 2012.

———. "Via rupta: vers la biodeconstruction." In *Appels de Jacques Derrida*, edited by D. Cohen-Levinas and G. Michaud. Paris: Hermann, 2014.

———. "With or Without You . . . Deconstructing Teleology between Philosophy and Biology." *Oxford Literary Review* 39, no. 1 (2017).

Weismann, August. *Essays upon Heredity*. Edited by Robert Robbins. Oxford: Clarendon Press, 1889.

———. *The Germ-Plasm, a Theory of Heredity*. Translated by W. Newton Parker and Harriet Rönfeldt. New York: Charles Scribner's Sons, 1893.

Wiener, Norbert. *The Human Use of Human Beings: Cybernetics and Society*. London: Free Association Books, 1989.

Index

Affirmation, 1, 186, 188–191, 194–195, 234. *See also* Attestation, Performative
Agapov, Eugene, 230–231
Aldrovandus, Ulysses, 56
Alterity, 4, 8, 27, 31, 80, 83, 92, 96, 106, 108–109, 120, 132–133, 170, 181, 183, 191, 195, 197, 223, 228–229, 233
Ameisen, Jean-Claude, 175–176, 180–182, 231, 233
Analogy, 23, 32–35, 59–63, 66, 73, 86, 90–91, 95, 97, 100–105, 108–113, 123–125, 140, 145, 148–151, 182, 198, 203, 208, 212, 221
Animal, 1–2, 20, 22–25, 34, 44–52, 56, 89, 106, 122, 124, 169, 205, 208, 219, 221
Apoptosis, 3, 151, 175, 233. *See also* Cellular suicide
Arche-writing, 7–11, 14–16, 18, 20, 24–27, 62–63, 73, 89–90, 106, 109, 111, 120, 122, 125, 128, 179, 188, 197, 201–202, 206, 234–235. *See also* Iterability, Reserve, Trace, Writing
Arche-trace, 202, 234. *See also* Trace
Aristophanes, 151–152
Aristotle, 49, 53–54, 56, 81–83, 104
Atlan, Henry, 223–224, 231

Attestation, 184, 189–191, 196–197, 199. *See also* Affirmation, Performative, Testimony
Austin, John L., 235
Autobiography, 169
Autoimmunity, 1, 3, 17, 151, 165, 167–176, 183–184, 228–231, 233

Bacteria, 91–99, 123–124, 217, 231–232
Bataille, George, 4–5, 40–42
Benjamin, Walter, 164
Bennington, Geoffrey, vii, 207, 212, 224–225, 227
Bergson, Henry, 208
Binding, 4, 5, 33, 45, 48, 130, 137–142, 147, 153–159, 162, 165, 197, 199, 209, 227–228. *See also* Bond
Biology, 1–4, 19–20, 27, 33–35, 41, 47, 49, 52–61, 65–66, 69, 73–79, 83, 90, 98, 103–105, 111, 113, 128, 147, 167, 174–177, 198, 203–208, 212–213, 216, 219, 221, 223, 226, 229–232
Biodeconstruction, 3–4, 7, 17, 202, 204
Biopolitics, 32, 207–208, 232–234
Biosemiotics, 218–219
Blanchot, Maurice, 1, 133, 187–190, 197

Boltzmann, Ludwig, 222
Bond, 61, 100, 127, 136, 191. See also Binding
Bonnet, Charles, 48
Borradori, Giovanna, vii
Breuer, Joseph, 137, 140
Brillouin, Léon, 123, 208

Canguilhem, George, 19, 29, 54–55, 61, 73, 104, 204, 211, 211
Cardan, Jérôme, 56
Carnot, Nicolas Léonard Sadi, 220
Cell, 12, 18, 27, 34–35, 48, 53, 55, 70–72, 79, 92, 96–101, 103–104, 107–108, 110, 123–124, 147–151, 170, 176–182, 203, 208–209, 217, 227, 230–231
Cellular suicide, 11, 151, 175–182, 233. See also Apoptosis
Certeau, Michel de, 187–188
Churilov, Leonid P., 230–231
Clausius, Rudolf, 220
Cohen Levinas, Danielle, vii
Community, 1, 26, 42, 167, 172–173, 183–184, 207, 235
Consciousness, 7–11, 16, 21, 24, 26, 40–42, 61, 68–69, 89, 121, 136, 153, 158, 170, 172–173, 183, 186, 188, 190, 192, 194–195, 202, 205, 235. See also Phenomenology
Constructivism, 64–65. See also Naturalism
Cybernetics, 2, 18–24, 27, 34–35, 54–55, 58, 60–61, 73, 103, 105, 112–115, 122–125, 203, 208, 213, 218, 221

D'Alambert, Jean-Baptiste Le Rond, 48
Death drive, 3, 127, 134–136, 142, 145, 147, 149–152, 155–156, 160–164, 170–171
Debru, Claude, 204

Deconstruction, 1–5, 7, 18, 29–33, 36, 49, 52–55, 58, 60, 66, 73–74, 92, 108, 112–113, 117, 122, 126, 128, 168, 176, 185–186, 206, 208–209, 212–213, 228, 234
Democracy, 3, 165, 167, 170–172, 183
Derrida, Marguerite, vii
Descartes, René, 211–212
Detour (*Aufschub*), 16, 129, 131–132, 143–144, 147, 163–164, 203, 209, 224–226
Determinism, 63–65, 69–73, 205–206, 208, 215, 218
Diderot, Denis, 48
Différance, 1, 4, 16–17, 19–22, 25–27, 29, 51–52, 62–64, 69, 73, 89–90, 105, 122, 128–136, 139, 141, 143–146, 154, 158, 162–165, 168, 172–173, 185, 193–198, 202–205, 207–208, 212, 215–216, 224, 228, 234
Dissemination, 50, 73, 98, 213, 235
DNA, 20, 34–35, 55, 98–99, 103–105, 204, 206, 209, 214, 217–218, 232
Doyle, Richard, 205–206
Drive of the Proper, 142–145, 152
Drive for Power, 152, 158–165, 170, 227–228
Duclaux, Emile, 230

Entropy, 113–115, 118–119, 123, 208, 220–222
Ermeneutics, 106
Epigenetics, 70–71, 212
Epistemology, 74–75, 108–111, 115, 211
Esposito, Roberto, 32, 207n8, 230n14, 232–234
Evolution, 1, 3, 9–13, 16–17, 20–26, 34, 48, 54, 59, 62–64, 66, 70, 89, 91–95, 99, 106, 120–122, 128, 135–136, 142–144, 148–149, 176, 204–205, 210, 212, 217, 219

Facioni, Silvano, vii, 201
Facilitation (*Bahnung*), 14–15, 120, 136–137
Favareau, Donald, 219
Fechner, Gustav Theodor, 127
Feedbeck regulation, 115, 118–121
Fernel, Jean François, 56
Foster, Michael, 201
Foucault, Michel, 208–209
Francis, Richard C., 71–73
Frauenfelder, Raoul, vii
Freud, Sigmund, 3, 5, 10–17, 30, 50, 65–66, 89–90, 121–122, 127–165, 170, 197, 201–203, 207, 216, 224–228, 235

Gadamer, Hans Georg, 106
Galenus, 56
Galilei, Galileo, 211
Garrido, Juan Manuel, vii
Gasché Rodolphe, vii, 203, 212, 218
Genetics, 19, 53, 65, 83, 107, 110–111, 147, 208–209, 213, 228
Genetic code, 33, 54, 56, 61, 67–69, 73, 83–84, 103–105, 198, 206, 209, 213
Goethe, Johann Wolfgang von, 49, 127, 226
Goette, Alexander Wilhelm, 128
Goldschmit, Marc, vii
Gramma/Grammé, 2, 18–22, 25, 27, 205–206

Hägglund, Martin, vii, 202, 208, 228, 229
Hartsoeker, Nicolas, 48
Hegel, Georg Wilhelm Friedrich, 2, 4, 20, 30–31, 33–56, 61, 76, 79–80, 82–84, 92, 144–145, 156, 210–211, 214
Heidegger, Martin, 29–30, 75, 106, 186, 188, 207, 207, 213

Helmoltz, Hermann von, 137
Heredity, 34, 54–59, 62, 66–67, 99, 103–104, 148, 198, 206, 208, 210, 226. See also Inheritance
Hering, Ewald, 128
Hippocrates, 56
History of Life, 9–10, 18, 21–22, 25–26, 62, 64, 205–206
HIV, 168–169, 176
Hyppolite, Jean, 19
Human being, 25–27, 45, 48, 50, 54, 63, 71, 78–79, 81, 88–90, 93, 104, 106, 123, 146, 186, 203, 205, 221–222, 224. See also Humanity
Humanism, 63–64, 73, 209
Humanity, 25, 68, 205. See also Human being
Husserl, Edmund, 2, 7–11, 14–16, 153, 170, 201–202, 214, 223, 229, 234

Immunitarian/ Immunity System, 4, 167, 169, 174–183, 229–233
Immunity, 167–168, 173–174, 180, 183–184, 207, 232–233
Immuno-depressant, 167, 174, 231
Inheritance, 59, 71–73, 98, 148, 151, 165, 212. See also Heredity
Information, 18–19, 23, 26–27, 35, 53–56, 59–61, 71, 79, 87, 104–106, 113–118, 122–123, 177–179, 198, 203–204, 206, 208, 211, 213, 215, 217, 221–223, 230–231
Iterability, 8–9, 15, 20, 24–27, 89, 106, 121, 128, 154, 173, 182, 191, 195–197, 204. See also Archewriting, Trace, Writing

Jablonka, Eva, 212
Jacob, François, 2, 3, 5, 19, 20, 29, 30, 33–35, 41, 52, 54–63, 66–128, 144–149, 198, 204, 208–226, 235
Jakobson, Roman, 218

Jämsä, Tuomo, 219
Johnson, Christopher, 203, 223
Johnson, David E., vii

Kamuf, Peggy, vii
Kant, Immanuel, 48, 56, 214
Kay, Lily E., 204
Kojève, Alexandre, 40–42, 210
Kull, Kalevi, 219

Lam, Marion, 212
Laplanche, Jean, 137
Leibniz, Gottfried Wilhelm von, 48, 82, 83
Leroi-Gourhan, André, 3, 20–26, 34, 63, 204–205
Lévi-Strauss, Claude, 26, 204, 213
Life-death, 1–4, 17, 33, 35, 122, 132–133, 143, 146, 165, 168, 171–173, 208, 225, 228–229, 234
Literature, 1, 127, 185, 187–191, 234–235
Living Present, 7–8, 14, 25, 121, 190–192, 196–199, 234. *See also* Phenomenology, Protention, Retention
Logic, 31–37, 41, 46, 50, 52, 58–63, 70, 78–79, 82–85, 87–90, 93–100, 103–106, 109, 121–125, 128, 134–135, 147, 167–170, 173–174, 183–184, 195, 197–198, 206–207, 211, 227–229, 234
Logos, 21, 27, 53–54, 60–63, 104, 212
Lotman, Yuri, 218
Low, Barbara, 152
Lwoff, André, 2, 19–20, 204

Machine, 76, 118, 122–125, 203, 208, 210, 212, 221, 231–232
Marx, Karl, 167, 173–174, 180, 183, 197

Malabou, Catherine, 73, 213
Mechanism, 12, 23, 34, 59, 69, 91, 98–99, 103, 118–119, 148, 167, 174, 208–211, 217, 230
Memory, 3–18, 21–34, 58–63, 66–69, 89, 95, 119, 136, 159, 169, 178–179, 194, 197–198, 202, 207
Mendoza de Jesus, Ronald, vii
Message, 33–35, 54–57, 61, 66–67, 103–106, 110, 115–126, 177, 198, 208, 221–222
Metaphysics, 2, 18–19, 31, 35–36, 60, 63, 73–74, 85, 87, 92, 116, 122, 172, 188, 190–191, 199, 206, 208, 213–215, 219, 223. *See also* Opposition
Mercier, Thomas Clément, vii
Metapsychology, 66, 130, 136
Model, 2, 18, 56, 66, 90, 95–97, 103–105, 108–117, 122–126, 137, 151, 171, 219, 230
Moles, Abraham, 19
Monod, Jacques, 2, 19, 20, 71, 204, 211–215
Moore, Gerald, vii
Morange, Michel, 204
Morin, Edgar, 224

Naas, Michael, vii, 231, 232
Narra, Hema Prasad, 98–99, 217
Naturalism, 64–65. *See also* Constructivism
Nielsen, Kaare M., 217
Nietzsche, Friedrich, 29–30, 83, 127, 207–208, 213

Ocham, Howard, 98–99, 217
Opposition, 1–3, 15, 18–19, 22, 24–25, 29–33, 35–42, 49, 60–68, 73–75, 85, 87, 92–99, 108, 110, 117–125, 132–134, 140–142, 146, 150–155, 172,

185–188, 195, 199, 205, 207–208, 213, 215–216, 222, 224, 227, 229, 233, 235. See also Metaphysics
Organization, 2, 17, 19, 26, 66, 77, 79, 90, 98, 101, 103, 106, 114, 116–125, 160–161, 173, 175, 178, 183, 209, 213, 224

Peeters, Benoît, 204
Peirce, Charles Sanders, 218–219
Performative, 4, 163, 185–199, 235. See also Affirmation
Phenomenology, 7–10, 14, 16, 24, 42, 120, 154–155, 186, 190, 223. See also Arche-writing, Consciousness, Living present, Retention, Protention
Philosophy of Life, 2, 20, 52, 54–55, 61, 73, 208
Physics, 43, 56, 75, 77, 79, 211, 214
Plato, 25, 60–61, 92, 127, 151–152, 212, 223
Poletaev, Alexander B., 230–231
Preformationism, 47–49, 71–72, 116
Program/Programme, 18, 20–24, 27, 34–35, 42, 53–59, 61, 63–75, 80, 83, 91, 94–101–104, 107, 110, 113, 118, 122, 147–148, 176–178, 198, 202, 205–206, 208, 211, 213–214
Protention, 21, 24, 106, 196–197. See also Living present, Phenomenology, Retention
Psychic system, 2, 10–17, 27, 63, 89, 121–122, 135–140, 153, 152, 156, 158, 165, 170
Psychic life, 14, 65, 127–129, 132, 139–143, 153, 228

Reductionism, 111
Reproduction, 4, 23–25, 34, 41, 48–52, 56–58, 67–69, 78–101, 103, 105, 108–110, 121, 124–125, 128, 148–149, 151, 160, 168, 172, 176, 181, 195, 198, 208–210, 217, 221
Reserve (*Vorrat*), 13, 16, 129, 202. See also Arche-writing
Retention, 7–11, 14–16, 21, 24, 89, 106, 120–121, 153, 190, 198. See also Living present, Phenomenology, Protention
Rückert, Friedrich, 127

Saussure, Ferdinand de, 25, 61
Schelley, Percy B., 191
Scherrington, Charles, 201n12
Schiller, Friedrich, 127, 226
Schopenhauer, Arthur, 127, 150
Schrödinger, Erwin, 123, 208, 220–222
Science, 11, 19–20, 30–35, 49, 52, 54–56, 68–71, 74–83, 87, 90, 97, 104, 107, 110–112, 122, 189, 198, 204, 208–209, 211–212, 214, 228–229, 232–233
Sebeok, Thomas E., 218–219
Senatore, Mauro, 4, 213
Sexuality, 12, 49–52, 67, 91–99, 120, 128, 131, 147–150, 158–160, 209, 217, 226
Sovereignty, 1, 3, 5, 75, 128, 155, 163–165, 167, 206, 228
Spallanzani, Lazzaro, 48
Spinoza, Baruch, 82, 83
Stiegler, Bernard, vii, 205
Stirner, Max, 173–174, 180, 183
Stricture, 133, 136, 140–141, 154–155, 158, 162, 197, 208, 211, 227
Stroev, Yuri I., 230–231
Structuralism, 116–117, 155, 204, 213, 223
Sublation (*Aufhebung*), 31, 35, 37–45, 47, 49–51, 79, 193

Supplementarity, 15, 32, 65, 73, 91, 93–95, 98–101, 120–121, 126, 144, 147, 149, 154–155, 175, 194, 197

Survival, 1, 11, 13, 16, 4, 17, 22–27, 33, 40, 69, 99, 106–108, 212, 132–133, 148, 165, 175–176, 181, 187, 192–220, 229, 231, 234–235

Swammerdam, Jan, 48

Technics/Technology, 3, 11, 25–26, 65, 87, 90, 115, 124, 173, 205, 220

Teleology, 53, 53, 57–58, 64, 73, 81, 83, 112, 124, 152, 156, 211–212, 214–216, 235

Teleonomy, 211, 214–215

Telic structure, 195–196, 199

Testament, 1, 4, 186, 196

Testimony, 1, 4, 189, 191, 195–197, 235. *See also* Attestation, Performative

Textuality, 2–5, 10–11, 17, 19, 27, 30, 32–35, 42, 67–68, 90, 96–101, 103–116, 120–126, 128, 154, 158, 169, 171, 196, 198, 216, 218–220, 228

Thermodynamics, 77, 113, 127, 137, 220, 222, 225

Thomas, Christopher M., 217

Trace, 1–11, 14–27, 46, 65, 73, 77, 89–90, 105–108, 120–121, 128–129, 136–137, 153–154, 165, 169–175, 180, 182, 186–192, 195–198, 202, 204, 206, 213, 218, 235. *See also* Arche-trace, Arche-writing, Iterability

Translation, 20, 57, 61, 104–106, 109–110, 126, 148, 150, 155, 159, 165, 169, 177, 196

Transmission, 7, 57–62, 74, 104–105, 118, 148, 165, 196, 198, 211, 215, 221, 223

Treviranus, Ludolf Christian, 49

Uexküll, Jacob von, 219

Vignola, Paolo, vii
Virchow, Rudolf, 230n14
Vitale, Francesco, 201, 202, 212, 216, 235

Weismann, August Friedrich Leopold, 128, 147–150

Wiener, Norbert, 19, 34, 103, 122–125, 203, 208, 221, 224

Writing, 2–5, 7, 10–11, 15, 18–22, 25–27, 31–32, 35, 53–55, 62–63, 65, 74–75, 90, 101, 104, 108, 112, 169, 185, 187, 195–198, 204, 206, 213, 215, 219, 232. *See also* Arche-writing, Iterability, Trace

www.ingramcontent.com/pod-product-compliance
Ingram Content Group UK Ltd.
Pitfield, Milton Keynes, MK11 3LW, UK
UKHW041917140426
5217IPUK00013B/195